The *Mathematica*® Programmer

LIMITED WARRANTY AND DISCLAIMER OF LIABILITY

AP PROFESSIONAL (APP) AND ANYONE ELSE WHO HAS BEEN INVOLVED IN THE CREATION OR PRODUCTION OF THE ACCOMPANYING SOFTWARE AND MANUAL (THE "PRODUCT") CANNOT AND DO NOT WARRANT THE PERFORMANCE OR RESULTS THAT MAY BE OBTAINED BY USING THE PRODUCT. THE PRODUCT IS SOLD "AS IS" WITHOUT WARRANTY OF ANY KIND (EXCEPT AS HEREAFTER DESCRIBED), EITHER EXPRESSED OR IMPLIED, INCLUDING, BUT NOT LIMITED TO, ANY WARRANTY OF PERFORMANCE OR ANY IMPLIED WARRANTY OF MERCHANTABILITY OR FITNESS FOR ANY PARTICULAR PURPOSE. APP WARRANTS ONLY THAT THE MAGNETIC DISKETTE(S) ON WHICH THE SOFTWARE PROGRAM IS RECORDED IS FREE FROM DEFECTS IN MATERIAL AND FAULTY WORKMANSHIP UNDER NORMAL USE AND SERVICE FOR A PERIOD OF NINETY (90) DAYS FROM THE DATE THE PRODUCT IS DELIVERED. THE PURCHASER'S SOLE AND EXCLUSIVE REMEDY IN THE EVENT OF A DEFECT IS EXPRESSLY LIMITED TO EITHER REPLACEMENT OF THE DISKETTE(S) OR REFUND OF THE PURCHASE PRICE, AT APP'S SOLE DISCRETION.

IN NO EVENT, WHETHER AS A RESULT OF BREACH OF CONTRACT, WARRANTY OR TORT (INCLUDING NEGLIGENCE), WILL APP BE LIABLE TO PURCHASER FOR ANY DAMAGES, INCLUDING ANY LOST PROFITS, LOST SAVINGS OR OTHER INCIDENTAL OR CONSEQUENTIAL DAMAGES ARISING OUT OF THE USE OR INABILITY TO USE THE PRODUCT OR ANY MODIFICATIONS THEREOF, OR DUE TO THE CONTENTS OF THE SOFTWARE PROGRAM, EVEN IF APP HAS BEEN ADVISED OF THE POSSIBILITY OF SUCH DAMAGES, OR FOR ANY CLAIM BY ANY OTHER PARTY.

Any request for replacement of a defective diskette must be postage prepaid and must be accompanied by the original defective diskette, your mailing address and telephone number, and proof of date of purchase and purchase price. Send such requests, stating the nature of the problem, to Academic Press Customer Service, 6277 Sea Harbor Drive, Orlando, FL 32887. APP shall have no obligation to refund the purchase price or to replace a diskette based on claims of defects in the nature or operation of the Product.

Some states do not allow limitation on how long an implied warranty lasts, nor exclusions or limitations of incidental or consequential damages, so the above limitations and exclusions may not apply to you. This Warranty gives you specific legal rights, and you may also have other rights which vary from jurisdiction to jurisdiction.

THE RE-EXPORT OF UNITED STATES ORIGIN SOFTWARE IS SUBJECT TO THE UNITED STATES LAWS UNDER THE EXPORT ADMINISTRATION ACT OF 1969 AS AMENDED. ANY FURTHER SALE OF THE PRODUCT SHALL BE IN COMPLIANCE WITH THE UNITED STATES DEPARTMENT OF COMMERCE ADMINISTRATION REGULATIONS. COMPLIANCE WITH SUCH REGULATIONS IS YOUR RESPONSIBILITY AND NOT THE RESPONSIBILITY OF APP.

The *Mathematica*® Programmer

ROMAN E. MAEDER

Department of Computer Science
Swiss Federal Institute of Technology
Zurich, Switzerland

AP PROFESSIONAL

A Division of Harcourt Brace & Company

Boston   San Diego   New York
London   Sydney   Tokyo   Toronto

Portions of this material have been adapted and revised for reuse with permission
from *The Mathematica Journal*, Miller Freeman, Inc. San Francisco, CA.

AP PROFESSIONAL
955 Massachusetts Avenue, Cambridge, MA 02139

An Imprint of ACADEMIC PRESS, INC.
A Division of HARCOURT BRACE & COMPANY

United Kingdom Edition published by
ACADEMIC PRESS LIMITED
24–28 Oval Road, London NW1 7DX

Library of Congress Cataloging-in-Publication Data
Maeder, Roman.
    The Mathematica programmer / Roman E. Maeder.
       p.   cm.
    Includes bibliographical references and index.
    ISBN 0-12-464990-4 (alk. paper)
      1. Mathematica (Computer program language) I. Title.
QA76.73.M29M33   1994
510'.285'53—dc20   93-34701
                               CIP

Printed in the United States of America
93 94 95 96 97 EB 9 8 7 6 5 4 3 2 1

Contents

Part 1: Paradigms of Programming

5 Databases

Part 2: Applications

6 Computations with Infinite Structures

7 Fibonacci on the Fast Track

8 Fractal Curves

9 Minimal Surfaces

10 System Programming

Appendices

Foreword

There's no need for us to worry so much about our computers any more. Thirty years ago, when there were at most a few computers in every city, there was little choice but to make things as easy for one's computer as possible. And that was the overriding concern in the design of early computer languages such as FORTRAN.

But now things should be different. There are more computers than people in the U.S., and almost every computer runs millions of instructions every second and can store millions of characters of data.

The strange thing is that most computers are still programmed in languages like FORTRAN. Despite all the computer resources that exist today, programmers still spend a lot of their time trying to make things as easy as possible for their computers.

This just doesn't make sense. What we should be doing is to use tools that make things as easy as possible for us, without worrying so much about our machines.

And indeed when I started the *Mathematica* project, this was one of the main ideas I had in mind. I wanted to build a computer language that was optimized for the way people think, not the way computers happen to be built.

In a sense, a computer language should be a precise codification of at least some limited aspects of human thinking. And in designing *Mathematica,* I tried to get inspiration from the various areas where such a codification had been attempted before — mathematics, philosophy, linguistics, some parts of computer science. And in the end, what we came up with was partly familiar, and partly radically new.

So how did it work out? I am happy to say that it has worked out very well. In the five years since *Mathematica* was released, more than a million people have used it, and have been exposed to the programming language ideas that it embodies.

But how can one really make use of these ideas? A common mistake is to focus only on those ideas that are familiar from some earlier programming experience, typically in a language like C or FORTRAN.

But learning *Mathematica* is a bit like learning a human language, and if you choose to use only those parts that you already know from another language, you will at best be speaking broken *Mathematica.*

But there is more than aesthetics at stake in using *Mathematica* fluently.

I consider myself an expert C programmer — I have written nearly a million lines of C code in my life. And I suppose it would be strange if I were not an expert *Mathematica* programmer.

Occasionally, I have the opportunity to write a program both in *Mathematica* and in C. I have a very simple observation about the comparison: the time it takes me to get the program working in *Mathematica* is usually at least a factor of ten less than in C.

And that kind of factor over and over again makes the difference between a project that can realistically be done, and one that cannot.

So why are people still using languages like C? Most of it, I believe, has nothing to do with technology. It has to do with issues of tradition, infrastructure, marketing, and so on.

It is not an easy matter to make a new programming language succeed. And in setting the strategy for *Mathematica,* we chose, from the name onward, to position the system initially as a tool for technical computing.

That has certainly worked out very well. But now it is time to take the computer language that is at the heart of *Mathematica,* and to understand how it can be used beyond the confines of technical computing.

There are exciting possibilities ahead. Indeed I suspect that within not too many years the idea of programming in a low level language like C will seem as specialized and esoteric as programming in microcode or assembler seems today.

So what is it really like to program in *Mathematica?* How can one get a feeling for the possibilities?

These are the kinds of questions that this book addresses. And in the capable hands of Roman Maeder—one of the original people I recruited to work on the *Mathematica* project—we get a first class tour of the world of *Mathematica* programming.

Stephen Wolfram

Preface

Mathematica's programming language is an important addition to a comprehensive symbolic computation system. To the novice, it may seem overwhelming, with close to 1000 built-in functions. As we will show in Chapter 1, it nevertheless has a coherent style and it is easy to learn. The significance of *Mathematica's* programming language can also be judged from the fact that it is used for teaching programming at a growing number of universities. One of the most rewarding features is that a clarifying picture and other means of explaining how programs work is never far away. One has the whole power of *Mathematica* at one's disposal.

The large number of ways to solve some given problem make it necessary to point out the easy way of doing things, especially to users who have been familiar with one of the traditional programming languages. The designers' view was expressed in my first book, *Programming in Mathematica.* My goal was to explain the ideas behind the language and to develop useful example programs. This strategy is continued in an ongoing series of articles in the *Mathematica Journal,* entitled *The Mathematica Programmer.* The title of this series is now also the title of this book. The articles can be divided roughly into two kinds: explanations of fundamental programming paradigms and applications. This distinction is reflected in the two parts that make up this book. Besides the articles from the journal, I also included new material, such as Chapter 1. The material from the articles has been expanded and all program listings are now included.

The first articles appeared before version 2.0 of *Mathematica.* Since there have been major improvements from version 1 to version 2, I took the opportunity to update all programs and their descriptions to version 2.

Pictures are always a good means of explaining things. The beauty of well-chosen graphics illustrations is in itself an important aspect of the otherwise rather dry world of computers. The color insert in this book gives me the chance to show the full range of possibilities. In all cases *Mathematica* delivered the raw data, which was then turned into color illustrations using a variety of techniques.

Our thanks go foremost to the past and present staff of the *Mathematica Journal:* Richard Rawles, Silvio Levy, Alan Zeichick, Troels Peterson, and Peter Altenberg.

We would like to thank *Miller-Freeman, Inc.,* for giving us permission to include the articles from the *Mathematica Journal* in this book.

Help with the color illustrations came from R. Peikert at IPS (Interdisciplinary Project Center for Supercomputing) of ETH, Zurich. We thank Stephen Wolfram for his inspiration and for the foreword. We are grateful to Chuck Glaser from AP Professional for encouraging us to publish this project and to Brian Miller for his production help. The expertise in phototypesetting at the AMS has greatly eased the burden of actually typesetting something like this.

Herrliberg, Switzerland   R. E. M.

About This Book

Contents of the Chapters

Part 1 discusses different important programming paradigms in greater detail. The introduction in Chapter 1 gives an overview. The next chapters treat abstract data types, polymorphism and message passing, object-oriented programming, and relational databases. The functional programming style is not treated in its own chapter, because it is pervasive in all our examples. Among the examples chosen for explaining the various topics are modular numbers, LISP-like programming, and collections (data types that can hold other data, such as lists, arrays, and dictionaries).

In Part 2 we show some interesting applications that make use of the ideas presented before. In Chapter 6 we work with infinite objects, such as lists and power series. Then come various programs for computing Fibonacci numbers. This gives us a good example for discussing optimizations and fast methods for computing powers of matrices. The next two chapters use *graphics*. First, we look at *fractal curves*. Besides the beautiful pictures we will learn a lot about interfacing our own data types with system functions. Then we will look at *minimal surfaces,* which gives us a chance to show how symbolic, numeric, and graphics methods can be combined to illustrate mathematical objects. A chapter on system programming concludes the main text of this book.

The appendix contains the bibliographic references and listings of programs not mentioned in the main text (especially the programs used to generate the illustrations in this book). This is followed by an index of all programs and the subject index.

An insert with color graphics contains high-resolution images related to the examples in the book, as well as some additional illustrations that show the range of possibilities. The programs for these illustrations are listed in the appendix and included on the disk, but not explained in the text.

This book is not an introduction to *Mathematica.* Syntactic details of its programming language are not treated here. *Mathematica* comes with an excellent manual (the *Mathematica* book) which explains everything we use here. The reference guide is now also available separately and is offered on-line on many computers (as a program called "MathBook").

About the Programs

All examples were tested with version 2.2 of *Mathematica.* The "live" calculation sequences in this book were computed on a Sun SPARCstation 2 running SunOS 4.1.1. No attempt has been made to ensure that programs work with earlier versions of *Mathematica.* Listing init.m shows the initialization commands used for the examples.

In the example *Mathematica* sessions in the later chapters, we will generally no longer show the command to read in the package that is the topic of the example. This command of the form

```
Format[Continuation[_]] := ""

SeedRandom[10000]

Off[ General::spell, General::spell1 ]

Begin["`Private`"]
Unprotect[Short]
Short[e_] := Short[e, 2]          (* lines are very short *)
Protect[Short]
End[]

Needs["SystemProgramming`"]

SetAllOptions[ColorOutput->GrayLevel] (* for book *)

SetOptions[ Plot3D, AspectRatio -> Automatic, PlotPoints -> 35 ]
SetOptions[ Graphics3D, AspectRatio -> Automatic ]
SetOptions[ ParametricPlot3D, Axes -> None ]

$DefaultFont = {"Times-Roman ISOLatin1", 10.0} (* ISO encoding for PS *)

SetOptions["stdout", PageWidth->60] (* line width *)
```

init.m: *Mathematica* initialization for this book

<<Package.m or *<<Package`* is assumed at the beginning of every session that uses functions
from the package in question. The latter form has the advantage that it knows how to convert the
context name into a file name valid on the given computer system. This is especially important
on old-fashioned systems that do not support longer file names.

Please observe that all examples have been computed in a fresh *Mathematica* session. While
the package mechanism minimizes the chance of a conflict between different functions it is an all
too common error to forget about global definition one has made earlier.

All programs mentioned are included in the floppy disk that comes with this book. It is a
standard 1.44MB MS-DOS 3.5in. floppy disk. Such disks can be read on the following
systems, among others: Sun SPARCstation, MS-DOS, Windows 3, Macintosh, and
NeXT. (Programs for reading MS-DOS floppies on the Macintosh are readily available.
We used *AccessPC* from Insignia Solutions.)
The disk contains directories for the various types of computers, named unix, pc, and mac. The
pc directory contains all files directly, with abbreviated names as is necessary under operating
systems such as MS-DOS. The other directories contain the files in compressed form, according
to system conventions. A plain text README file explains how they can be unpacked. This
preserves the original file names. We recommend that you copy the directory best suited for you
onto the hard disk and include its path name in *Mathematica's* search path $Path. This can be
done with a command

AppendTo[$Path, "*full pathname of directory*"]

in your init.m file. The form of the pathname is system dependent.

Notation and Terminology

Mathematica input and output is typeset in typewriter-like style: Expand[(x+y)∧9]. Parts of such input that are not literal, but denote (meta-)variables, are typeset in italics: f[*var_*] := *body*. *Functions* or *commands* are referred to by their name followed by an empty argument list, for example Expand[ ]. Listings of programs are delimited with horizontal lines and usually have captions beneath them.

Names of files containing programs are typeset in Helvetica typeface: ParametricPlot3D.m. Notebooks have an extension of .ma. Genuine dialogue with *Mathematica* is set in two columns. The left column contains explanations and the right column contains the input and output, including graphics. You should be familiar with these conventions from *Programming in Mathematica* and from the *Mathematica* book.

Colophon

The manuscript was written in LaTeX [26] (with many extensions). It contains only the input of the numerous computation examples. The results were computed by *Mathematica* and then automatically inserted into the input file for TeX. The bibliography was produced with BibTeX [39] and the index was sorted with the help of makeindex [27]. Figures not produced with *Mathematica* were generated with FRAMEMAKER on a NeXT computer and converted into POSTSCRIPT. The output of TeX was then combined with all POSTSCRIPT graphics, using dvips[41] and output on a phototypesetter.

The color images not produced directly from *Mathematica* were rendered with AVS [2] (a scientific visualization system) or RAYSHADE [25] (a ray tracing program) on Silicon Graphics computers at IPS, the Interdisciplinary Project Center for Supercomputing at ETH, Zurich.

---

[1] *MathSource* is an electronic archive of contributed *Mathematica* programs and packages maintained by Wolfram Research, Inc.. To find out more about it, send an e-mail message with the single line help intro to mathsource@wri.com.

Part 1
Paradigms of Programming

Chapter One
Introduction

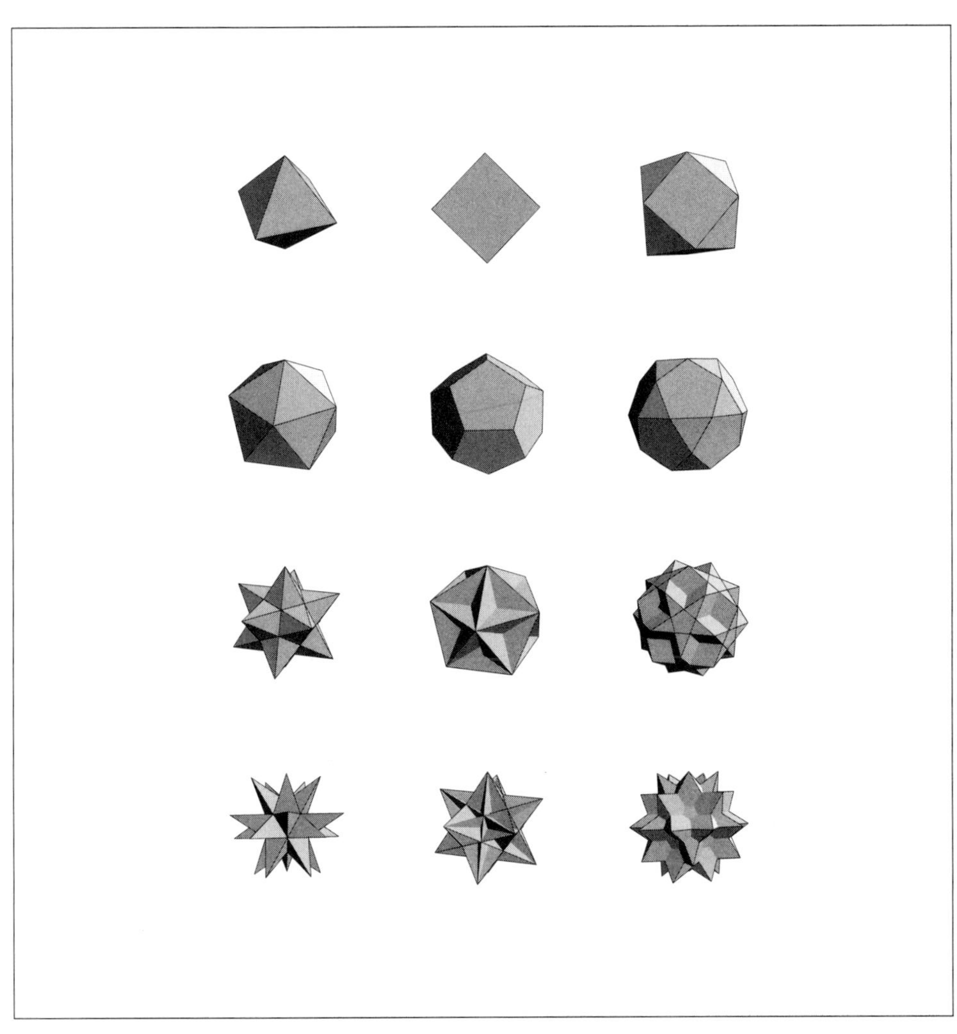

Mathematica includes a rich and powerful programming language. It combines the procedural, functional, and rule-based programming styles in a single coherent system. We describe the design decisions taken in implementing *Mathematica's* programming language and compare the different programming styles in the way they present themselves in *Mathematica*.

About the illustration overleaf:
Regular and semiregular polyhedra. For each pair of dual regular polyhedra there is a semiregular polyhedron, obtained from either of the two regular ones by cutting off the vertices. The first two rows show the ordinary (convex) Platonic solids, the next two rows show the four non-convex regular polyhedra (the Kepler–Poinsot solids). In each row, the associated semiregular solid is shown last.

From left to right and top to bottom:
octahedron, cube, cuboctahedron,
icosahedron, dodecahedron, icosidodecahedron,
small stellated dodecahedron, great dodecahedron, dodecadodecahedron,
great stellated dodecahedron, great icosahedron, great icosidodecahedron.

The code for this picture is in Pictures.m. It uses the package UniformPolyhedra.m from [31]. Some of these solids are also shown on Plate 3.

    The great icosidodecahedron is also the basis for the cover illustration. There it was rendered with ray tracing, which adds shadows and a more realistic lighting model.

1.1 *Mathematica's* Programming Language

Mathematica has been available since 1988 and is being used more and more in teaching, research, and in industry. A by-product of the symbolic computation system is a programming language that differs from traditional languages in many important ways. It combines ideas from many different sources.

- Conventional languages are not well suited to express mathematical formulae and algorithms. LISP and other functional languages showed alternatives.

- An important tool is the possibility to express mathematical rules in a simple way. This requires good pattern matching capabilities. Ideas were taken from PROLOG.

- Another prerequisite is the simple manipulation of structured data. This has been pioneered by APL.

- Modularization is an important tool for organizing larger programs.

- Object-oriented elements make code development easier. Ideas were taken from SIMULA, SMALLTALK, and C++.

- Traditional procedural programming in the style of PASCAL, FORTRAN, and **C** is also provided.

All of this taken together leads to a large language with many built-in functions. It nevertheless has a consistent and uniform style. This becomes possible by using only *rewrite rules* as the underlying mechanism for implementing all other programming constructs. Such a language is also interactive and therefore easy to use. It is not necessary to compile functions or to embed them into a main program to use them.

1.1.1 A Uniform Paradigm

Strictly speaking, there are no procedures, functions, or subroutines in *Mathematica*. Any definition of the form

$$f[args] := body$$

is a *rewrite rule*. Whenever the evaluator sees an expression that matches the left side, it is replaced by the right side with the values of the pattern variables substituted. This corresponds closely to a procedure call of a traditional language, a similarity that is intended. Definitions can therefore be used in place of traditional procedures, functions, and rewrite rules.

A procedure body typically declares some local variables and consists of a sequence of assignment statements and control structures.

```
SplitLine[vl_] :=
      Module[{vll, pos, linelist = {}, low, high},
             vll = If[NumberQ[#], #, Indeterminate]& /@ vl;
             pos = Flatten[ Position[vll, Indeterminate] ];
             pos = Union[ pos, {0, Length[vll]+1} ];
             Do[ low = pos[[i]]+1;
                 high = pos[[i+1]]-1;
                 If[ low < high,
                     AppendTo[linelist, Take[vll, {low, high}]] ],
                {i, 1, Length[pos]-1}];
             linelist
        ]
```

A typical procedure

A function contains nested calls of other functions and returns a result.

```
RandomPoly[x_, n_] :=
      Sum[ Random[Integer, {-10, 10}] x^i, {i, 0, n} ]
```

A typical function

A rewrite rule transforms an expression into another one.

```
log[a_ b_] := log[a] + log[b]
```

A typical rewrite rule

1.1.2　Parameters of Procedures

An important difference from traditional procedural languages concerns the treatment of *formal parameters* of functions or procedures. Superficially a definition in *Mathematica*, such as

$$f[x\_, \ y\_] := body$$

looks (intentionally) similar to a function declaration. The pattern variables in such a definition are, however, not local variables in the procedure (as they are in PASCAL or **C**)! The values of the arguments are *inserted* for every occurrence of the pattern variables in the body of the definition.

It is an instructive exercise to emulate the two standard parameter passing mechanisms of procedural languages: *call by value* and *call by reference*.

Parameter Passing by Value

It is easy to emulate **C**-style procedure parameters:

```
f[x0_, y0_, z0_] :=
        Module[ {x=x0, y=y0, z=z0},
                    ⋮
            ]
```
C-style procedure parameters

The idea is to use initialized local variables. The variables x, y, and z are then proper local variables. This code corresponds to the following **C** declaration:

```
f(x, y, z)
{
    ⋮
}
```
Function parameters in **C**

This function finds the fixed point of $x \mapsto 1 + \frac{1}{x}$ with starting value x_0.

```
In[1]:= fixedpoint[x0_] :=
            Module[{x = x0},
                While[ x != 1 + 1/x, x = 1 + 1/x ];
                x
            ]
```

The fixed point is the *Golden Ratio*.

```
In[2]:= fixedpoint[1.2]

Out[2]= 1.61803
```

Exercise: What happens if the pattern variable is named x_ and Module[ ] is left out?

Parameter Passing by Reference

On the other hand, parameter passing by reference (var parameters in PASCAL, pointers in **C**, references in C++) is also possible:

```
SetAttributes[f, HoldAll]

f[xref_Symbol] :=
        Module[{locals...},
                    ⋮
                xref = val (* assignment to global parameter *)
            ]
```
Parameter passing by reference

Since the attribute `HoldAll` prevents evaluation of the actual argument in a call of the function `f`, the unchanged arguments are inserted into the body of the function. Any computation done is therefore effectively performed on the argument itself. No other (local) variables are involved.

Attributes should be set before any rules are defined.	`In[3]:= SetAttributes[inc, HoldAll];`
This defines a function `inc` that increments its parameter.	`In[4]:= inc[nref_Symbol] := nref++`
Here is a global variable with value 5.	`In[5]:= var = 5;`
Since `inc` does not evaluate its argument, the operation is performed on the global variable `var`.	`In[6]:= inc[var]` `Out[6]= 5`
The value of `var` has indeed been changed.	`In[7]:= var` `Out[7]= 6`
It does not make sense to call `inc` with anything other than a symbol. The pattern does not match and nothing happens.	`In[8]:= inc[17]` `Out[8]= inc[17]`

Exercise: What happens if `inc` does not have the attribute `HoldAll`?

1.2 Pattern Matching and Term Rewriting

Pattern matching and term rewriting represent the fundamental principle of *Mathematica's* evaluator. All other programming constructs are implemented in terms of it. It is especially useful for implementing rules corresponding to transformations from a handbook of formulae. Equations in a handbook are usually meant to be used as rewrite rules, transforming the left-hand side into the right-hand side. By looking up an expression we essentially perform pattern matching in our head.

1.2.1 Example: Laplace Transforms

Definitions can be taken almost verbatim from a handbook of mathematics. Here is an excerpt from the standard package Calculus‘LaplaceTransforms‘. It shows some simple rules for Laplace transforms of polynomials and exponential functions. The first three definitions express the fact that the Laplace transform is a linear transform.

```
LaplaceTransform[c_, t_, s_] := c/s /; FreeQ[c, t]

LaplaceTransform[a_ + b_, t_, s_] := LaplaceTransform[a, t, s] + LaplaceTransform[b, t, s]

LaplaceTransform[c_ a_, t_, s_] := c LaplaceTransform[a, t, s] /; FreeQ[c, t]

LaplaceTransform[t_^n_., t_, s_] := n!/s^(n+1) /; (FreeQ[n, t] && n > 0)

LaplaceTransform[a_ t_^n_., t_, s_] :=
        (-1)^n D[LaplaceTransform[a, t, s], {s, n}] /; (FreeQ[n, t] && n > 0)

LaplaceTransform[a_. Exp[b_. + c_. t_], t_, s_] :=
        LaplaceTransform[a Exp[b], t, s-c] /; FreeQ[{b, c}, t]
```

Packages can be referred to by their context name. This is a machine independent specification.

```
In[1]:= << Calculus`LaplaceTransform`
```

Here is the transform of $e^{\omega t + \phi} t^2$.

```
In[2]:= LaplaceTransform[ Exp[omega t + phi] t^2, t, s]
```

$$Out[2]= \frac{2\,E^{phi}}{(-omega + s)^3}$$

1.2.2 Mathematical Programming

Mathematical concepts can easily be translated into programs. For example, the properties of the greatest common divisor (gcd) of two numbers

$$\begin{aligned}
\gcd(a, b) &= \gcd(b, a \bmod b) \\
\gcd(a, 0) &= a
\end{aligned}$$

which lead to Euclid's algorithm, can be turned into a running program almost verbatim:

```
gcd[a_, b_] := gcd[b, Mod[a, b]]]
gcd[a_, 0] := a
```

The definitions can immediately be used.

```
In[1]:= gcd[1999, 2999]
Out[1]= 1
```

The built-in tracing and debugging facilities make it easy to study how the program works. Here we see the sequence of calls of gcd.

```
In[2]:= Trace[ gcd[5, 8], gcd[__Integer] ] // TableForm
Out[2]//TableForm= gcd[5, 8]
                   gcd[8, 5]
                   gcd[5, 3]
                   gcd[3, 2]
                   gcd[2, 1]
                   gcd[1, 0]
```

1.3 Programming Styles

This section shows examples of some commonly identified programming styles.

1.3.1 Functional Programming

Given the problem of adding the square roots of the first 500 integers, the solution in most programming languages is to use an auxiliary variable that is incremented by the square root of a loop index iterating from 1 to 500.

```
sum = 0.0;
Do[ sum = sum + N[Sqrt[i]],
    {i, 1, 500} ];
sum
```

In symbolic computation systems, this loop reduces to a single statement

```
Sum[ N[Sqrt[i]], {i, 1, 500} ]
```

which corresponds directly to the mathematical formula

$$\sum_{i=1}^{500} \sqrt{i}\,. \tag{1.3--1}$$

Mathematica does not force one to think about *how* to implement a summation, but lets one focus on the *concept* itself.

Folding

What does `Fold[ ]` do? One easy way to see how such functions work is to apply them to purely symbolic (undefined) functions.

The binary operator `f` is applied to the result of the previous application and the next element from the list. `e0` is used for initializing this process.

```
In[1]:= Fold[f, e0, {e1, e2, e3}]

Out[1]= f[f[f[e0, e1], e2], e3]
```

One use of `Fold[]` is for *recursion removal*. The body of a recursively defined function (with one recursive call) can be thought of as a function f of two arguments: the result of the recursive call and the value of the recursion parameter (the value that changes from one call to the next). We then generate a list of all values of the recursion parameter and use `Fold[]` with the initial value equal to the value of the boundary case. Here is the standard example, factorial numbers. This is the recursive definition:

```
factorial1[0] = 1;
factorial1[n_] := n factorial1[n-1]
```

The body function is simply multiplication, the recursion parameter ranges from 1 to n. This gives the following definition:

```
factorial2[n_] := Fold[ Times, 1, Range[1, n] ]
```

Note that it works also for $n = 0$. `FoldList[]` instead of `Fold[]` shows one the intermediate steps.

Here are all factorials from 0 to 9.

```
In[2]:= FoldList[ Times, 1, Range[1, 9] ]
Out[2]= {1, 1, 2, 6, 24, 120, 720, 5040, 40320, 362880}
```

Lazy Evaluation

With lazy evaluation, arguments of functions are only evaluated when they are needed, not when the function is called. It can be implemented with the attributes `HoldFirst` and `HoldRest`. Among other things this idea allows us to implement *infinite lists*. We will do this in Chapter 6.

1.3.2   Structural Operations

Powerful operations allow us to treat large structured objects (vectors, matrices, nested lists of things) as single objects. This avoids most of the loops found in traditional programming languages. (It is interesting to note that FORTRAN, the language used most often for numerical computation, lacks operations on vectors and matrices as a whole. One always has to write a loop.)

Many combinatorial problems can be solved with structural operations.

Here is the list of all possibilities of choosing one of `a1`, `a2`, and `a3`, one `b` and one of `c1`, `c2`.

```
In[1]:= Distribute[{{a1, a2, a3}, b, {c1, c2}}, List]
Out[1]= {{a1, b, c1}, {a1, b, c2}, {a2, b, c1},
         {a2, b, c2}, {a3, b, c1}, {a3, b, c2}}
```

This is the list of all grid points next to the origin (in two dimensions).

```
In[2]:= Flatten[Outer[List, {-1,0,1}, {-1,0,1}], 1]

Out[2]= {{-1, -1}, {-1, 0}, {-1, 1}, {0, -1}, {0, 0},
         {0, 1}, {1, -1}, {1, 0}, {1, 1}}
```

This gets rid of the origin itself. What remains are all neighboring points.

```
In[3]:= neighbors = Complement[%, {{0, 0}}]

Out[3]= {{-1, -1}, {-1, 0}, {-1, 1}, {0, -1}, {0, 1},
         {1, -1}, {1, 0}, {1, 1}}
```

This definition chooses a random direction each time the symbol `randDir` is evaluated.

```
In[4]:= randDir :=
            neighbors[[ Random[Integer, Length[neighbors]] ]]
```

How do we find the permutation that sorts a list into standard order?

Here is a sample list.

```
In[1]:= l = {1.2, 5.4, -2.1, 33, 18.5};
```

We attach the index of each number in a list of pairs.

```
In[2]:= Transpose[{l, Range[Length[l]]}]

Out[2]= {{1.2, 1}, {5.4, 2}, {-2.1, 3}, {33, 4}, {18.5, 5}}
```

Now we can sort it.

```
In[3]:= Sort[ % ]

Out[3]= {{-2.1, 3}, {1.2, 1}, {5.4, 2}, {18.5, 5}, {33, 4}}
```

We keep only the indices. They describe the original position of the numbers, or the necessary permutation.

```
In[4]:= tr = #[[2]]& /@ %

Out[4]= {3, 1, 2, 5, 4}
```

The permutation can be used to rearrange other lists of dependent information into the same order.

```
In[5]:= {a, b, c, d, e}[[tr]]

Out[5]= {c, a, b, e, d}
```

1.3.3 Logic Programming

Logic programming or *declarative programming* tries to write declarations that express certain properties of the desired result without specifying the flow of control. Pattern matching and backtracking are then used to solve an instance of the problem.

As an example, here is in typical PROLOG style a program for reversing a list.

```
reverse[l_] := rev[ l, {} ]
rev[ {}, r_ ] := r
rev[ {h_, t___}, {r___} ] := rev[ {t}, {h, r} ]
```

The auxiliary function `rev[`*left*, *right*`]` operates on two stacks (lists). It removes the first element from the left list and puts it onto the right list.

A trace shows how this code works. It uses the method we humans would use to reverse the order of a deck of cards: taking one after another and placing it onto a second pile.

```
In[1]:= TracePrint[ reverse[{a, b, c, d, e}], rev[_,_] ]

rev[{a, b, c, d, e}, {}]
rev[{b, c, d, e}, {a}]
rev[{c, d, e}, {b, a}]
rev[{d, e}, {c, b, a}]
rev[{e}, {d, c, b, a}]
rev[{}, {e, d, c, b, a}]

Out[1]= {e, d, c, b, a}
```

Backtracking can be implemented with side conditions. The pattern matcher generates all possible cases. The side condition can then be used to commit a certain case.

An *inversion* in a list is a pair of adjacent elements such that the first one is larger than the second one. A sorted list is characterized by not having any inversions of adjacent elements. To sort a list, we simply reverse any inversions found. No particular order of doing this needs to be specified.

```
sort[ {alpha___, x_, y_, omega___} ] := sort[ {alpha, y, x, omega} ] /; x > y

sort[ l_ ] := l
```

The pattern matcher generates all possible pairs of adjacent x and y. Whenever they are out of order, they are reversed.

```
In[2]:= sort[ {5, 1, 3, 2} ]

Out[2]= {1, 2, 3, 5}
```

1.3.4 Abstract Data Types

Abstract data types are both a theoretically well-defined concept and a useful tool for program development. Following the principles of abstract data type design, one arrives at a clear separation of *specification* and *implementation*.

Abstract data types are defined in terms of type names, function names, and equations. These can be realized in *Mathematica* very easily. The equations become rewrite rules. The interactive nature of *Mathematica* makes it well suited for rapid prototyping and testing of designs.

We will talk about abstract data types in Chapter 2.

1.3.5 Object-Oriented Programming

Object-oriented programming is a programming style that is becoming more and more popular. It promises code-reuse and easier maintenance of larger projects than is possible with traditional procedural languages. Its use of methods and message passing instead of procedure calls shifts the programmer's view toward close integration of data and operations.

Objects and Classes

The first important aspect of object-oriented languages is that functions are considered part of data. A data object "knows" which operations can be performed on it. The functions defined for a certain type of object are part of that object and are called *methods*.

Methods are usually not defined for each object separately but are collected in a class. Objects then belong to a class from which they take their methods.

Inheritance

Often a number of related data types have some common characteristics. Some operations on them can be written in a way that does not depend on which of the data types it is applied to. Common characteristics of related data types can then be isolated and encapsulated in a new data type. The other data types are then made subtypes of the new type. They inherit the characteristics of the common type and need only implement those aspects in which they differ from their supertype.

An interactive object-oriented language can easily be implemented in *Mathematica*. We will do this in Chapter 4.

1.3.6 Modularization

Modern programming languages provide features for organizing large programming projects. Most important are *modularization* (or encapsulation) and *information hiding*. *Mathematica* calls such programs *packages*. A package consists of two parts: an *interface definition* and an *implementation part*. The interface describes the aspects that a client needs to know. The implementation provides that functionality but it is hidden from users of the package. Here is how such a package looks like in *Mathematica* (this example is taken from [29]):

```
BeginPackage["Reshape`"]

Reshape::usage = "Reshape[list, dims] rearranges list
    as a nested structure with dimensions dims."

Begin["`Private`"]

reshape[list_, {n_}] := list /; Length[list] == n
reshape[list_, {head__, n_}] :=
    reshape[ Partition[list, n], {head} ]
Reshape[list_, dims_] := reshape[ Flatten[list], dims ]

End[]
Protect[ Reshape ]
EndPackage[]
```

The part between the initial `BeginPackage[]` and `Begin["`Private`"]` is the interface. It declares all functions exported from this package (there is only one: `Reshape`) and documents them. The part between `Begin[]` and `End[]` is the implementation. The auxiliary function `reshape` is not exported. It will not be available to users of this package. The command `Protect[Reshape]` prevents any modification of the exported function `Reshape` by users of our package.

Chapter Two
Abstract Data Types

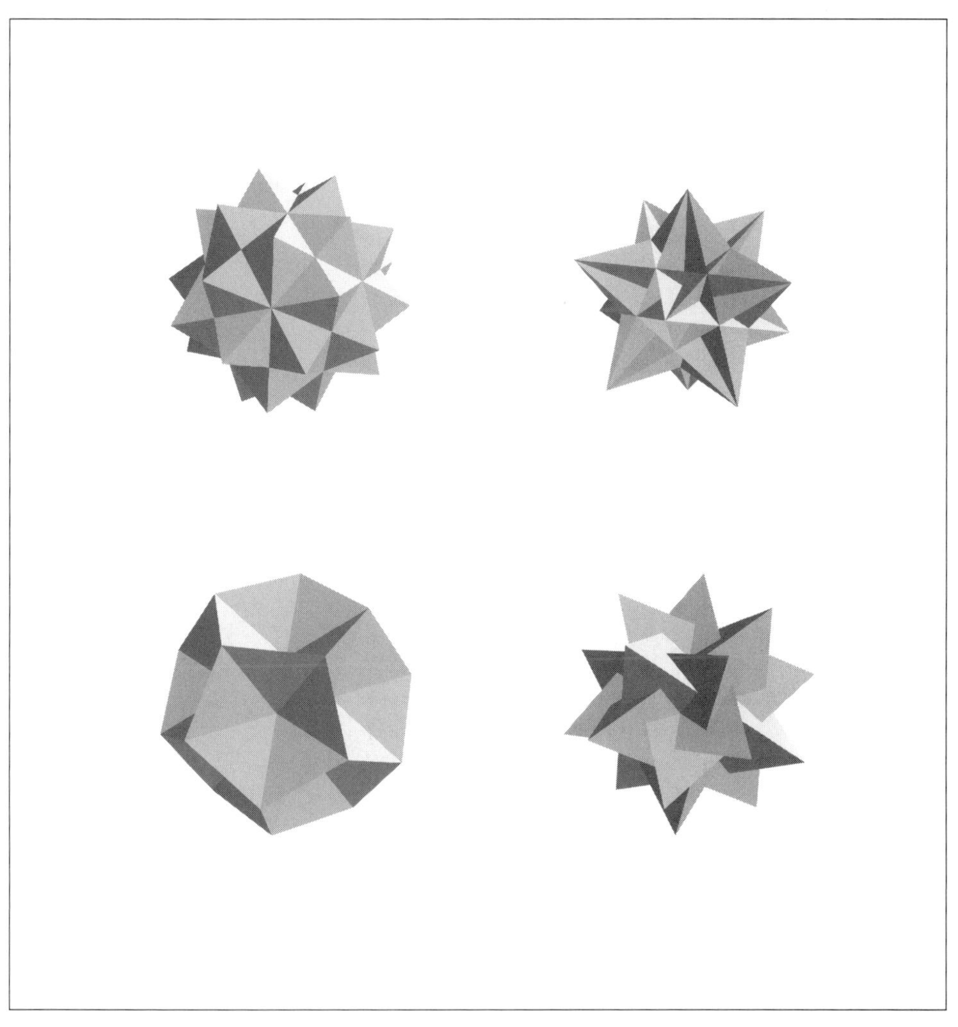

The theoretical concept of abstract data types provides a useful tool for program development. Following the principles of abstract data type design ensures a clear separation of specification and implementation. Abstract data types are defined in terms of type names, function names, and equations. These can be realized in *Mathematica* very easily. The equations become rewrite rules. *Mathematica's* interactive nature makes it well suited for rapid prototyping and testing of designs.

About the illustration overleaf:
Four of the 59 stellations of the icosahedron.
upper left: No. 3, the compound of five octahedra,
upper right: No. 11, the great icosahedron (see also the illustration at the beginning of Chapter 1),
lower left: No. 20,
lower right: No. 36, the compound of five tetrahedra.

The code for producing all 59 forms is in the package Icosahedra.m. The enumeration of all stellations was accomplished by H. S. M. Coxeter [9]. A gallery of 12 of these stellations is reproduced in Plate 6.

2.1 Data Types as a Programming Tool

Texts on programming emphasize to a varying degree the importance of abstraction, that is, the capturing of common concepts from a variety of concrete manifestations. The concept of a data type is one such abstraction. A data type can be defined roughly as a set of values and the methods to access them. One can then decide on data representations and the functions with which data are manipulated and follow a usually self-imposed discipline in using only these functions in the application program. We will arrive at such guidelines through a study of the theoretical foundations of the design of abstract data types. Usually, such theoretical foundations are of, well, theoretical interest only since they are either nonconstructive to begin with or cannot be made into executable programs on a real computer. The specification of an abstract data type can be made executable under certain conditions, however. This is particularly easy in *Mathematica*, since it contains an interactive symbolic programming language with rewrite rules. Therefore, please bear with me during the following rather theoretical definitions of an abstract data type. We will turn those into running *Mathematica* programs as soon as possible.

2.2 The Definition of Abstract Data Types

An *algebra* is a mathematical structure that consists of a set (of elements) and a number of operations on these elements. The integers with the usual arithmetic operations are an example. Since we often use more than one type of data, we need to expand this notion into *many-sorted* algebras. The task of defining the operations then becomes a bit tricky, since we have to specify the type of all operands.

An *abstract data type* is specified in terms of three parts:

- A set S of *sort names.*

- A set Σ of *function names* (or operator symbols).

- A set E of *equations.*

The triplet $\langle S, \Sigma, E \rangle$ is called a *specification.* The sorts can be thought of as the names of the "types" we are going to model. Usually there is more than one; we may use auxiliary types, such as **bool**, standing for Boolean values (truth values). We will print sort names in boldface, all lowercase.

In our first example **Nat1**, we will model the non-negative integers, with Boolean values as an auxiliary type. Therefore S consists of two elements:

$$S = \{\textbf{int}, \textbf{bool}\} \, . \tag{2.2--1}$$

The function names are broken down according to their "arity," that is the number and types of arguments and the return type. The arity is described by a sequence of elements from S, the last one being the return type, the other ones describing the argument types in the proper order. Formally, the set Σ consists of sets $\Sigma_{w,s}$, with $s \in S, w \in S^*$. (This says that w is a sequence of elements of S, and s is a single element of S, the return type). Elements of $\Sigma_{\lambda,s}$ are called *constants* since they are functions without arguments (λ denotes the empty sequence).

Continuing our example, here are a few commonly defined function names:

$$\begin{aligned}
\Sigma_{\mathbf{int}} &= \{\mathbf{z}\} \\
\Sigma_{\mathbf{bool}} &= \{\mathbf{f}, \mathbf{t}\} \\
\Sigma_{\mathbf{int\ int}} &= \{\mathbf{s}\} \\
\Sigma_{\mathbf{int\ bool}} &= \{\mathbf{isz}\} \\
\Sigma_{\mathbf{bool\ bool}} &= \{\mathbf{not}\} \\
\Sigma_{\mathbf{int\ int\ int}} &= \{\mathbf{add}, \mathbf{mult}\}
\end{aligned} \tag{2.2--2}$$

This says, for example, that $\mathbf{z}$ is a constant of type **int** and that **add** is a function of two arguments of type **int**, returning a result also of type **int**. Σ is called the *signature*. Please note that we have not given these function names any meaning yet, even though eventually we will of course think of **add** as the addition function on integers.

There is nothing in the definitions so far that forces us to use the addition function as meaning for the function named **add**. To make our intention clear, we can give a number of equations that must be satisfied by the functions we choose to implement the operations in Σ. To write down equations, we will use the operation symbols themselves as purely formal functions together with variable symbols for each sort. Here are the equations that enforce the usual meaning for our operations. Variables are printed in italics. Their type is obvious from the functions in which they appear. If this were not the case we would have to define the types of the variables formally.

$$\begin{aligned}
\mathbf{isz}(\mathbf{z}) &= \mathbf{t} \\
\mathbf{isz}(\mathbf{s}(n)) &= \mathbf{f} \\
\mathbf{not}(\mathbf{t}) &= \mathbf{f} \\
\mathbf{not}(\mathbf{f}) &= \mathbf{t} \\
\mathbf{add}(n, \mathbf{z}) &= n \\
\mathbf{add}(n, \mathbf{s}(m)) &= \mathbf{s}(\mathbf{add}(n, m)) \\
\mathbf{mult}(n, \mathbf{z}) &= \mathbf{z} \\
\mathbf{mult}(n, \mathbf{s}(m)) &= \mathbf{add}(\mathbf{mult}(n, m), n)
\end{aligned} \tag{2.2--3}$$

We are now ready to define Σ-*algebras*. A Σ-algebra is a mathematical structure that "fits" the signature Σ. It consists of *sets* for the sorts from S and of mappings corresponding to the operator symbols from Σ.

An S-*sorted* Σ-*algebra* $\mathbf{B}$ consists of a set B_s for each $s \in S$, called the *carrier set* of sort s, and a function

$$\sigma_B : B_{s_1} \times B_{s_2} \times \cdots \times B_{s_n} \to B_s \tag{2.2--4}$$

for each $\sigma \in \Sigma_{w,s}$, with $w = s_1 s_2 \ldots s_n$ (i. e., a function taking arguments from the carrier sets corresponding to its arity).

We can now endow our example **Nat1** with its intended meaning, using the non-negative integers as carrier set of sort **int**. To describe the functions we use *Mathematica* notation.

$$
\begin{aligned}
B_{\textbf{int}} &= \{0, 1, 2, \ldots\} \\
B_{\textbf{bool}} &= \{\texttt{False}, \texttt{True}\} \\[4pt]
\mathbf{z}_B &= \texttt{0\&} \\
\mathbf{f}_B &= \texttt{False\&} \\
\mathbf{t}_B &= \texttt{True\&} \\
\mathbf{s}_B &= \texttt{Function[n, n + 1]} \\
\mathbf{isz}_B &= \texttt{Function[n, n == 0]} \\
\mathbf{not}_B &= \texttt{Not} \\
\mathbf{add}_B &= \texttt{Plus} \\
\mathbf{mult}_B &= \texttt{Times}
\end{aligned}
\qquad (2.2\text{--}5)
$$

It is customary to leave out the empty argument sequence for constants. Instead of writing (`False&`)`[]` we will simply write `False` from now on.

An S-sorted Σ-algebra **B** *satisfies the equations* if for each substitution of elements of the appropriate sorts for all variables in an equation the left and right side of the equation become equal. A Σ-algebra satisfying the equations E is called a Σ, E-algebra. One can prove by induction that the algebra **B** given earlier satisfies the equations of **Nat1**.

2.2.1 The Typical, or Initial, Σ-Algebra

A Σ, E-algebra does not yet capture our idea of a "data type." There are usually several different (non-isomorphic) Σ-algebras having additional properties (not required by the equations E) and it is not clear which one is the "right one." For example, here is another algebra satisfying **Nat1**.

$$
\begin{aligned}
B_{\textbf{int}} &= \{0, 1\} \\
B_{\textbf{bool}} &= \{\texttt{False}, \texttt{True}\} \\[4pt]
\mathbf{z}_B &= \texttt{0\&} \\
\mathbf{f}_B &= \texttt{False\&} \\
\mathbf{t}_B &= \texttt{True\&} \\
\mathbf{s}_B &= \texttt{Function[n, 1]} \\
\mathbf{isz}_B &= \texttt{Function[n, n == 0]} \\
\mathbf{not}_B &= \texttt{Not} \\
\mathbf{add}_B &= \texttt{Max} \\
\mathbf{mult}_B &= \texttt{Min}
\end{aligned}
\qquad (2.2\text{--}6)
$$

Because the carrier set $B_{\textbf{int}}$ has only two elements, it is easy to verify that the equations are satisfied. Clearly this is not what we intended the natural numbers to be.

There is, however, one special algebra that satisfies the equations and has no other spurious properties. It is named $T_{\Sigma,E}$ and is constructed from the *term algebra* T_Σ by identifying any terms that are forced equal by the equations. The carrier sets of the term algebra T_Σ consist of terms built by using the operator symbols as functions, just as in the equations given earlier. The elements of the carrier sets of $T_{\Sigma,E}$ are equivalence classes of elements of T_Σ.

The details of this abstract construction can be found in the literature about abstract data types [15, 14]. This algebra is called the *initial* Σ, E-algebra. (It is the initial object in the category of all Σ, E-algebras.) It is this algebra that captures our notion of abstract data type.

If the equations are sufficiently well behaved, we can model this algebra in *Mathematica*. In our example the equations can be thought of as *reductions,* always replacing the left side by the (simpler) right side. In this way there is one special term in each class of terms identified by the equations. It is the term that is in normal form with respect to the reductions derived from the equations. Any other term is transformed into normal form by the applicable reduction rules.

Not every specification is executable in this sense. An equational theory need not be decidable (a famous undecidable theory is the theory of finitely presented groups, for example). If the theory is not decidable we cannot find a set of rewrite rules with the property that any two terms in the same equivalence class reduce to the same term and that all reductions terminate in a finite number of steps. In our simple example **Nat1** we can simply turn the equations into rewrite rules. These rules turn out to be sufficient for reducing all equivalent terms into normal form. Usually the rewrite rules derived from the equations are not sufficient for this purpose since equations allow us to go back and forth, while rules are always applied left-to-right. If the original rules are not sufficient it may be possible to find additional rules. These additional rules can be found in a mechanical way by the Knuth-Bendix algorithm [23]. This algorithm finds the missing rules. If it terminates we get out of it a set of rewrite rules sufficient to decide equivalence of terms. Since there are undecidable theories, it may never terminate, however.

The term algebra T_Σ can be modeled in *Mathematica* by using the operator symbols as unspecified functions. To implement $T_{\Sigma,E}$ we define the equations as rewrite rules. Any term not in normal form will be simplified into normal form by these rules as we have seen. This technique is called *executable specification.* Listing Nat1.m shows the code for **Nat1**. Note that we turn the variables in the equations into "named blanks" in the rules.

Here are a few sample computations with **Nat1**. The integer n is represented as n-fold nesting of **s** applied to **z**.

<table>
<tr><td>This performs the addition $1 + 2 = 3$.</td><td>

```
In[1]:= add[ s[z], s[s[z]] ]
Out[1]= s[s[s[z]]]
```

</td></tr>
<tr><td>A trace shows the computations going on.</td><td>

```
In[2]:= Trace[ add[ s[z], s[s[z]] ], add,
               TraceForward->True]
Out[2]= {add[s[z], s[s[z]]], s[add[s[z], s[z]]],
        {add[s[z], s[z]], s[add[s[z], z]], {add[s[z], z], s[z]},
        s[s[z]]}, s[s[s[z]]]}
```

</td></tr>
</table>

```
(* the initial algebra for the specification Nat1 *)

(* sorts *)

int
bool

(* Sigma *)

{z}              (* -> int *)
{t, f}           (* -> bool *)
{s}              (* int -> int *)
{isz}            (* int -> bool *)
{not}            (* bool -> bool *)
{add, mult}      (* int int -> int *)


(* equations turned into rewrite rules *)

isz[z] := t
isz[s[n_]] := f

not[t] := f
not[f] := t

add[n_, z] := n
add[n_, s[m_]] := s[add[n, m]]

mult[n_, z] := z
mult[n_, s[m_]] := add[mult[n, m], n]
```

Nat1.m: The abstract data type **Nat1** in *Mathematica*

The product of 2 and 3.	`In[3]:= mult[ s[s[z]], s[s[s[z]]] ]`
	`Out[3]= s[s[s[s[s[s[z]]]]]]`
This is the test 1 =! 0, returning true.	`In[4]:= not[isz[s[z]]]`
	`Out[4]= t`

2.3   A Practical Approach

You wouldn't want to perform arithmetic with this abstract version of integers. We can work with a more suitable Σ, E-algebra as long as it is isomorphic to $T_{\Sigma,E}$. We have already given such an implementation of **Nat1** in terms of *Mathematica*'s built-in integers and arithmetic operations. One can show that our implementation of **Nat1** is isomorphic to the abstract term algebra given above. The integers $0, 1, 2, \ldots$ are in one-to-one correspondence with the terms 0, s[0], s[s[0]], $\ldots$.

Still, this purely formal approach is of great value. It allows the formulation of the intended behavior of a data type without referring to its implementation. Any implementation can later be used for the actual computation. Moreover if all application code uses exclusively the operator symbols from Σ the implementation can be changed without changing any application code! This is probably the most important aspect of using abstract data types. It allows a clear separation between specification and implementation. Often there is no need to go all the way to a formal specification $\langle S, \Sigma, E \rangle$, but it suffices to classify the functions into *constructors, selectors, constants,* and *predicates.*

Consider the example **Mod**, intended to model $\mathbf{Z}_p$, the ring of integers modulo some positive integer p. We can build on top of **Nat1**, extending S by a sort **mod**, and adding the following operations:

$$
\begin{aligned}
\Sigma_{\textbf{int}} &= \{\mathbf{p}\} \\
\Sigma_{\textbf{mod int}} &= \{\mathbf{rep}\} \\
\Sigma_{\textbf{int mod}} &= \{\mathbf{makemod}\} \\
\Sigma_{\textbf{mod mod mod}} &= \{\mathbf{add}, \mathbf{mult}\} \\
\Sigma_{\textbf{int int int}} &= \{\mathbf{rem}\}
\end{aligned}
\qquad (2.3\text{--}1)
$$

There is no problem with using the same operation name (here **add** and **mult**) in different sets $\Sigma_{w,s}$. They are distinguished by the type of arguments and return values. **rep** is meant to give a *representative* of a modular number, an integer that can be used to implement arithmetic in **mod**, as we will see in the following equations. **rem** is the integer remainder function, whose definition is omitted. It is needed only to express the desired properties of **rep** in the equations. **makemod** turns an integer into a modular number. Here are the equations:

$$
\begin{aligned}
\mathbf{makemod}(\mathbf{rep}(m)) &= m \\
\mathbf{rem}(n - \mathbf{rep}(\mathbf{makemod}(n)), \mathbf{p}) &= \mathbf{0} \\
\mathbf{add}(m_1, m_2) &= \mathbf{makemod}(\mathbf{add}(\mathbf{rep}(m_1), \mathbf{rep}(m_2))) \\
\mathbf{mult}(m_1, m_2) &= \mathbf{makemod}(\mathbf{mult}(\mathbf{rep}(m_1), \mathbf{rep}(m_2)))
\end{aligned}
\qquad (2.3\text{--}2)
$$

The second equation says that the representative of a modular number made from an integer n is congruent to n modulo p ("$-$" being integer subtraction). This is usually expressed as

$$
\mathbf{rep}(\mathbf{makemod}(n)) \equiv n \quad (\mathrm{mod}\ p). \qquad (2.3\text{--}3)
$$

Addition and multiplication are then defined in terms of their integer counterparts through the use of **rep**.

Here is one possible implementation, using expressions of the form mod[n], n an integer, to hold elements of $B_{\textbf{mod}}$.

```
makemod[n_Integer] := mod[ Mod[n, p] ]
rep[ mod[n_] ] := n

mod/: m1_mod + m2_mod := makemod[ rep[m1] + rep[m2] ]
mod/: m1_mod * m2_mod := makemod[ rep[m1] * rep[m2] ]
```

Mod0.m: Modular numbers as abstract data type

The use of the built-in function $\texttt{Mod}[n, p]$ in **makemod** ensures that a unique representative is chosen (an integer in the range $0 \ldots p - 1$). Putting this computation into the constructor removes all worries about the choice of representative from the rest of the code and trivially satisfies the first and second equation.

Note that we have *overloaded* the built-in addition and subtraction functions. This corresponds directly to the use of the same names in the specification.

This sets the modulus to 5 for the computations that follow.	`In[1]:= p = 5`
	
	`Out[1]= 5`

This makes two modular numbers and assigns them to the two variables m2 and m3.	`In[2]:= m2 = makemod[2]; m3 = makemod[3]`
	
	`Out[2]= mod[3]`

Their sum is 0 (modulo 5).	`In[3]:= m2 + m3`
	
	`Out[3]= mod[0]`

Their product is 1 (modulo 5).	`In[4]:= m2 m3`
	
	`Out[4]= mod[1]`

Our implementation is not yet complete. There are no definitions for subtraction of modular numbers, for example.	`In[5]:= m2 - m3`
	
	`Out[5]= mod[2] - mod[3]`

We call **makemod** a *constructor,* because it creates values of a certain type (here **mod**). **rep** is a *selector,* since it returns parts of an expression (**mod** elements have only one part, the representative of the modular number). Later on we will also encounter *predicates.* These are functions that return a Boolean value.

2.4 Design Principles for Abstract Data Types

From the previous example, we can derive the general recommended way of defining and implementing an abstract data type.

How to define a data type:

- Define the names of the sorts to be used.
- Define the constants to be used.
- Define constructors, selectors, and predicates.
- Write down the equations that should hold.

How to implement a data type:

- Choose a representation for elements of the sorts defined.
  Usually you can use a normal expression having the name of the sort as head. This corresponds roughly to a record in many programming languages.

- Derive rules for normal forms for the data elements.
  This ensures that elements are stored in a unique way. The rules are derived from the equations and can normally be put into the constructors.

- Define constructors, selectors, and predicates.
  These are the only operations that are allowed access to the details of the internal representations.

- Define the other operations.
  An operation for which exactly one equation exists can usually be turned into a simple rewrite rule. (This is called a *derived* operation.) The variables in the left side of the rules are in the form n_type, restricting the arguments of the functions implemented to the correct types. Use selectors only to access parts of the data elements on the right side of the rules.

- Use overloading where appropriate.
  Overloading built-in functions is most easily done by defining upvalues, rules of the form $g/:\ f[n\_g,\ldots]\ :=\ \ldots$.

- Choose a suitable output representation of data elements.
  If you do this, you can consider declaring the sorts themselves in the private implementation part of your package. Users are then prevented from accessing them directly.

- Define automatic type conversions where useful.
  This is a convenience only.

In addition, the framework of a package should be used, with documentation for the exported functions in the usual way. Listing Modular.m is a sample package for computation with modular numbers that shows some of the preceding ideas, continuing the example **Mod**.

This call defines to modulus to use.	`In[1]:= SetModulus[17]`
	`Out[1]= 17`
The constant 1 mod 17 is handy to have around.	`In[2]:= one = MakeModular[1]`
	`Out[2]= 1 mod 17`
Any integer can be turned into a modular number by multiplying it with `one`.	`In[3]:= 100 one`
	`Out[3]= 15 mod 17`
Large powers can be computed efficiently using the built-in function `PowerMod[ ]`.	`In[4]:= % ^ 99999999999999999999999999999`
	`Out[4]= 8 mod 17`

```
BeginPackage["Modular`"]

SetModulus::usage = "SetModulus[p] sets the modulus to be used."
MakeModular::usage = "MakeModular[n] returns n mod p."
Representative::usage = "Representative[m] returns a representative
    of the modular number m."

Begin["`Private`"]

`theModulus        (* a private static variable *)
`mod               (* the private name of the data type *)

(* constructors *)

MakeModular[n_Integer] := mod[ Mod[n, theModulus] ]
SetModulus[p_Integer?Positive] := (theModulus = p)

(* Selectors *)

Representative[mod[n_]] := n

(* Arithmetic *)

mod/: a_mod + b_mod := MakeModular[ Representative[a] + Representative[b] ]
mod/: a_mod * b_mod := MakeModular[ Representative[a] * Representative[b] ]
mod/: a_mod ^ q_Integer :=
    MakeModular[ PowerMod[Representative[a], q, theModulus] ]

mod/: a_mod + b_Integer := a + MakeModular[b]
mod/: a_mod * b_Integer := a * MakeModular[b]

(* Output Formatting *)

Format[m_mod] := SequenceForm[Representative[m], " mod ", theModulus]

End[ ]

EndPackage[ ]
```

Modular.m: A package for modular numbers

Our rules also work for subtraction.

```
In[5]:= 1 - %
Out[5]= 10 mod 17
```

Division is turned into multiplication and a negative power. For a modular number a, a^{-1} is the *modular inverse*.

```
In[6]:= 1/%
Out[6]= 12 mod 17
```

The product of a number and its inverse should be one.

```
In[7]:= % %%
Out[7]= 1 mod 17
```

This small loop computes the *order* of the modular number 2. The order is the smallest exponent n, such that $x^n = 1 \quad (\text{mod } p)$.

```
In[8]:= x = MakeModular[2]; n = 1;\
        While[ x^n =!= one, n++]; n
Out[8]= 8
```

2.5   An Example: LISP in *Mathematica*

The data structure of the programming language LISP can easily be modeled in *Mathematica*. LISP expressions consist of atoms and dotted pairs, written

$$(e_1 . e_2) \, , \tag{2.5–1}$$

where e_1 and e_2 are again expressions. The function $\mathrm{cons}(e_1, e_2)$ creates the dotted pair $(e_1 . e_2)$, $\mathrm{car}(pair)$ returns the first element of a pair, and $\mathrm{cdr}(pair)$ returns the second element. Atoms are symbols, numbers, and strings. We treat all *Mathematica* atoms as LISP atoms. The expression

$$(e_1 . (e_2 . (\dots e_n . \mathrm{nil}) \dots)) \tag{2.5–2}$$

is called a list and is written

$$(e_1 \ e_2 \ \dots \ e_n) \tag{2.5–3}$$

(with a blank instead of a comma between the elements!) The empty list $()$ is equal to the atom `nil`. LISP function calls are written as lists, with the function being the first element and the argument following. For example: `(cons a b)`, `(car e)`. Here is the abstract data type of LISP, in *Mathematica* notation.

- Atoms: Numbers, symbols, strings (all *Mathematica* atoms).

- Constants: `nil` and all atoms.
  Note that this gives an infinite number of constants. Nothing in the theory of abstract data types demands a finite number of operation symbols.

- Constructors: `cons[`e_1`, `e_2`]`, `list[`e_1`, `e_2`, ..., `e_n`]`, for $n \geq 0$.
  (`list` stands for an infinite number of operator symbols, one for each number of arguments.) `list` is not a fundamental constructor, but a very convenient derived operator that makes it easy to build lists.

- Selectors: `car[`e`]`, `cdr[`e`]`.

- Predicates: `pairQ[`e`]`, `atomQ[`e`]`, `nullQ[`e`]`.
  `nullQ` is true for the empty list and false for other lists.

- Equations:
  The first set of equations expresses the basic relationship between the selectors `car` and `cdr` and the constructor `cons`.

$$\begin{aligned} \mathrm{car}[\mathrm{cons}[e_1, e_2]] &= e_1 \\ \mathrm{cdr}[\mathrm{cons}[e_1, e_2]] &= e_2 \end{aligned} \tag{2.5–4}$$

The following equations define atoms, pairs, and empty lists. Note that the number of equations is infinite. There is one equation for atomQ for each atom a.

$$
\begin{aligned}
\texttt{atomQ[a]} &= \texttt{True} \quad \text{(one equation for each atom a)} \\
\texttt{atomQ[cons[}e_1\texttt{, }e_2\texttt{]]} &= \texttt{False} \\
\texttt{pairQ[}e\texttt{]} &= \texttt{!atomQ[}e\texttt{]} \\
\texttt{nullQ[nil]} &= \texttt{True} \\
\texttt{nullQ[cons[}e_1\texttt{, }e_2\texttt{]]} &= \texttt{False}
\end{aligned}
\tag{2.5--5}
$$

The constructor list is a derived operation. There is one equation for each $n \geq 0$.

$$
\begin{aligned}
\texttt{list[]} &= \texttt{nil} \\
\texttt{list[}e_1\texttt{, }e_2\texttt{, ..., }e_n\texttt{]} &= \texttt{cons[}e_1\texttt{, list[}e_2\texttt{, ..., }e_n\texttt{]],} \quad n > 0
\end{aligned}
\tag{2.5--6}
$$

The implementation of this data type is simple. We use cons as head for the LISP cells (this corresponds to a constructor without any rules attached to it). The auxiliary function listToList is used to speed up the special output formatting that makes these lists print the way they do in LISP. See Listing Lisp.m at the end of this section for the complete code.

Here is a sample LISP list.
```
In[1]:= t = list[a, b, c, d, e]
Out[1]= (a b c d e)
```

The car of a list is its first element.
```
In[2]:= car[t]
Out[2]= a
```

The cdr of a list is the *rest* of it, that is, the list without its first element.
```
In[3]:= cdr[t]
Out[3]= (b c d e)
```

This is a typical recursive LISP definition for appending an element to a list.
```
In[4]:= append[l_, e_] :=
            If[ nullQ[l], list[e],
                cons[car[l], append[cdr[l], e]] ]
```

This is a (slow) way of reversing a list.
```
In[5]:= reverse[l_] :=
            If[ nullQ[l], nil,
                append[reverse[cdr[l]], car[l]] ]
```

It works just as it does in LISP.
```
In[6]:= reverse[t]
Out[6]= (e d c b a)
```

In LISP, we cannot take the car of an atom. In our theory this leads merely to a term that cannot be simplified further since no equations apply to it.
```
In[7]:= car[atom]
Out[7]= car[atom]
```

We will look at a closely related data type, infinite lists, in Chapter 6.

```
BeginPackage["Lisp`"]
cons::usage = "cons[a, b] gives the dotted pair (a .b)."
car::usage = "car[pair] gives the first element of pair."
cdr::usage = "cdr[pair] gives the second element of pair."
list::usage = "list[e1, e2,..., en] gives the Lisp list (e1 e2 ... en)."

nil::usage = "nil is the empty Lisp list."

pairQ::usage = "pairQ[e] is true, if e is a dotted pair."
atomQ::usage = "atomQ[e] is true, if e is an atom."
nullQ::usage = "nullQ[l] is true, if l is the empty list."
Begin["`Private`"]

$RecursionLimit = Infinity (* necessary for long lists *)

(* constructors *)

list[] = nil
list[e_, r___] := cons[e, list[r]]

(* selectors *)

car[cons[e_, l_]] := e
cdr[cons[e_, l_]] := l

(* predicates *)

atomQ[_?AtomQ] = True
atomQ[_cons] = False     (* leave undefined otherwise *)
pairQ[e_] := !atomQ[e]
nullQ[nil] = True
nullQ[_cons] = False

(* output formats *)

listToNest[nil] := {}
listToNest[c_cons] := {car[c], listToNest[cdr[c]]}
listToList[l_] := Flatten[listToNest[l]]

listQ[nil] = True        (* this atom is also a list *)
listQ[l_cons] := listQ[cdr[l]]
listQ[_] = False

Format[nil] = "()"
Format[l_cons?listQ] := SequenceForm["(", Infix[listToList[l], " "], ")"]
Format[l_cons] := SequenceForm["(", Infix[l, " . "], ")"]

(* fix Infix problem for functions with one argument *)

protected = Unprotect[Infix]
Infix[_[e_], h_:Null] := e
Protect[Evaluate[protected]]

End[]
Protect[ car, cdr, cons, pairQ, atomQ, nullQ, list ]
EndPackage[]
```

Lisp.m: LISP in *Mathematica*

2.6 Built-in Functions and Data Types

In Chapter 8 we present a data type for *fractal curves*. We define a data type **curve** and then overload Show[] so that such a curve can be plotted as if it were one of the existing graphics types, such as Graphics, SurfaceGraphics, etc. This is an important concept. It allows the use of pre-existing functions for new user-defined data types. In this way one doesn't have to learn new function names but can use the already familiar ones, for example Show to display a graphics object, whether built-in or user-defined. Many sections in our "Programming in *Mathematica*" are devoted to the methods of how to make user-defined data types behave like built-in ones.

Another example of the power of abstraction is *sorting:* We can write a general-purpose sorting routine that merely needs an ordering predicate as an argument. It doesn't need to know anything else about the data to be sorted. Formally, such an ordering predicate is in $\Sigma_{\textbf{data data bool}}$, where **data** is whatever data type we want to order. The predicate returns true if its first argument is to be considered less than or equal to the second one. *Mathematica*'s built-in sorting procedure Sort[*list, p*] takes advantage of this by allowing you to provide your own ordering predicate *p* (the default amounts to the canonical ordering). The behavior of Sort described is an example of *polymorphism.* A polymorphic function can work on data of many different types.

2.7 An Exercise

Database operations can also be viewed as instances of abstract data types. This simple example maintains a database of *records,* with functions for insertion, retrieval, and deletion. Note that the records themselves are treated as purely abstract. As in the sorting example mentioned earlier, we need to know very little about them in order to put them into a database. Records are assumed to contain a *key field* and there is an ordering predicate defined on the possible values of this key field.

The implementation of the database is also left open. You can choose the implementation, be it a (possibly) sorted list, a binary tree, or some hashing scheme. The choice depends on the anticipated frequency of the various operations to be performed. For simplicity you can assume that all key field values appear at most once. Here is the specification:

- Constants:
  emptyBase: the empty database.
  NullRecord: a value used to signal failure of a search.

- Constructors:
  Insert[*database, record*] inserts a record into a database. If a record with the same key as

the one to be inserted is already in the database, nothing is done.
`Drop[`*database*`, `*key*`]` removes the record with the given key field value from the database, if present.
Both constructors return the possibly modified database.

- Selectors:
  `Key[`*record*`]` returns the value of the key field of the record.
  `Retrieve[`*database*`, `*key*`]` returns the record with the given key field value from the database. If no such record exists, `NullRecord` is returned.

- Predicates:
  `KeyOrder[`*key$_1$*`, `*key$_2$*`]` returns -1 if $key_1 < key_2$, 0 if both keys are equal, and +1 if $key_1 > key_2$. (This is not strictly speaking a predicate, but easier to use than an ordering function that returns only two values.)
  `MemberQ[`*database*`, `*key*`]` returns true, if a record with the given key exists in the database.

You may want to write down the formal Σ-algebra specification for this. You need four sorts, one for databases, one for records, another one for the keys, and one for the values of the sorting function. Deriving the equations is not trivial. You will have to express the fact that inserting a record twice has no effect, for example. For more on databases see Chapter 5.

2.8   Conclusions and Further Reading

Abstract data types are one important method for clean program design. The theory of many-sorted algebras is not adequate for all purposes, however. For one, not all domains can be described by equations alone (fields, for example the rational numbers, are not an equational theory). We might also want to capture the notion of an "error" or illegal operation (such as division by zero). For these purposes, the simple notion of a carrier *set* is no longer sufficient. One needs mathematical models with more structure. One such model is the *universal domain,* which forms the basis for *denotational semantics.*

In Chapter 6 we give another example of a data type, one that represents infinite lists and overloads many built-in functions to work with infinite lists as if they were ordinary lists.

The two basic references for abstract data types have already been mentioned. A survey of the current state of the field can be found in Wirsing's article in the *Handbook of Theoretical Computer Science* [48]. The same handbook also contains an article on rewrite systems by Dershovitz and Jouannaud [10].

Any book on programming will emphasize the importance of a clear distinction between interface and implementation. We have said nothing new in this respect. For example, Abelson & Sussman [1] develop the notions of constructors, selectors, and predicates in order to implement data types in the language SCHEME, which is—like *Mathematica*—untyped to begin with. Fundamental references to denotational semantics include [42, 43]. **Nat1** is the familiar *Peano model* of arithmetic.

 The floppy disk contains the packages Mod0.m, Nat1.m, Modular.m, and Lisp.m.

Chapter Three
Polymorphism and Message Passing

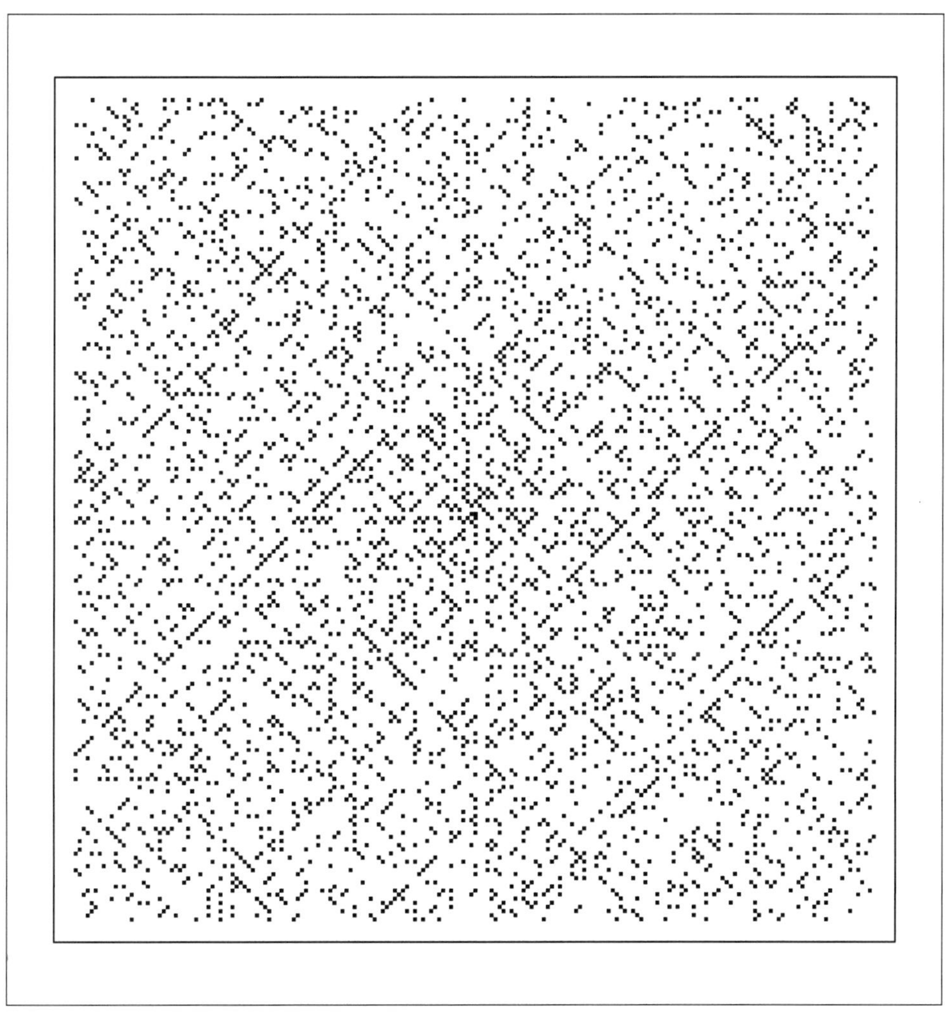

Polymorphic functions are functions that can handle several different types of arguments. We look at ways of implementing such functions. First we present three LISP-like implementations using dispatch tables, data-driven programming, and message passing. Then we look at various ways of implementing objects, that is, data elements that have built in all the functions that can be applied to them. This concept provides for a high degree of modularity and lays the groundwork for a discussion of object-oriented programming in Chapter 4.

About the illustration overleaf:
The location of primes. The integers are represented in a spiral fashion (with the number one in the center) and primes are marked black. The visible diagonal lines represent certain arithmetic progressions containing many primes. The picture was produced with

```
PrimesPlot[99],
```

using the package Spirals.m. See also Pictures.m. A similar plot showing the number of divisors of integers is given in Plate 1.

3.1 Polymorphic Operations

One tool for writing "reusable" software is *polymorphism*. Often, the same function needs to be applied to data of different types. Rather than write separate functions for each data type, we try to write one function that can handle data of different types. We will look at several ways to implement this idea.

We will explain our ideas with an example: polymorphic arithmetic with complex numbers. We will represent complex numbers in two different ways: Cartesian (rectangular) coordinates and polar coordinates. The arithmetic functions must therefore allow for arguments of either of these types. Note that we do not work with *Mathematica*'s built-in complex numbers in this example. Instead, we will implement complex numbers as *abstract data types*. Here is the formal definition of the two data types for rectangular and polar complex numbers:

Constructors	`makeRectangular[x, y]`	make the number $x + iy$
	`makePolar[r, phi]`	make the number $re^{i\phi}$
Predicates	`rectangularQ[z]`	is z in rectangular coordinates?
	`polarQ[z]`	is z in polar coordinates?
Selectors	`realPartRectangular[z]`	the real part Re z
	`realPartPolar[z]`	
	`imagPartRectangular[z]`	the imaginary part Im z
	`imagPartPolar[z]`	
	`magnitudeRectangular[z]`	the magnitude $|z|$
	`magnitudePolar[z]`	
	`angleRectangular[z]`	the phase angle arg z
	`anglePolar[z]`	

Each of these "primitive" (nonpolymorphic) operators works with numbers of a particular type (rectangular or polar). We will look at several ways to develop the following polymorphic selectors and predicates, which should operate on numbers of either type:

`realPart[z]`	the real part Re z
`imagPart[z]`	the imaginary part Im z
`magnitude[z]`	the magnitude $|z|$
`angle[z]`	the phase angle arg z
`complexQ[z]`	is z a complex number?

Using these selectors and predicates, we can define the polymorphic arithmetic operations listed in CmplxA1.m. These definitions will stay the same for the different implementations of the data types that we will present.

To implement the four basic arithmetic operations we need rules for addition and multiplication, a rule for the negative of a number, and one for its inverse. The rule for inverse is generalized

```
unprotected = Unprotect[Plus, Times, Power]

z1_?complexQ + z2_?complexQ :=
    makeRectangular[ realPart[z1] + realPart[z2],
              imagPart[z1] + imagPart[z2] ]

-z_?complexQ := makeRectangular[ -realPart[z], -imagPart[z] ]

z1_?complexQ * z2_?complexQ :=
    makePolar[ magnitude[z1] * magnitude[z2],
              angle[z1] + angle[z2] ]

(z_?complexQ)^n_Integer :=
    makePolar[ magnitude[z]^n, n angle[z] ]

Protect[ Evaluate[unprotected] ] (*restore protection *)
```

CmplxA1.m

to allow for any integer power, not just z^{-1}. *Mathematica* turns subtraction into addition and negation, and division into multiplication and inversion, therefore we do not need additional rules for subtraction and division.

A difference is turned into a sum and negation.	```In[1]:= FullForm[a - b]``` ```Out[1]//FullForm= Plus[a, Times[-1, b]]```
A quotient is turned into a product and inverse.	```In[2]:= FullForm[a / b]``` ```Out[2]//FullForm= Times[a, Power[b, -1]]```

3.2   Three Implementation Methods

Somewhere in the implementation of the polymorphic operators the data type of the argument needs to be determined to decide which primitive operator to call. The three methods are distinguished by *where* this decision is made.

3.2.1   Dispatch Tables

One way of implementing the polymorphic selectors is to dispatch on the type of the operand by testing all possible cases (there are only two in our example). (A dispatch table is a device to select one of several alternative pieces of code depending on the value of a variable or the outcome of a sequence of tests. In *Mathematica*, these can be implemented with Switch and Which, respectively.) The code is in the program CmplxDis.m.

   It is now straightforward to implement the two underlying data types. We use the symbols rectangular and polar as heads of expressions holding the two coordinates of the complex

```
(* Dispatch Tables for polymorphic Selectors *)

complex::badtype = "`1` is not a complex number."

realPart[z_] :=
    Which[ rectangularQ[z], realPartRectangular[z],
           polarQ[z],       realPartPolar[z],
           True,            Message[complex::badtype, z] ]

imagPart[z_] :=
    Which[ rectangularQ[z], imagPartRectangular[z],
           polarQ[z],       imagPartPolar[z],
           True,            Message[complex::badtype, z] ]

magnitude[z_] :=
    Which[ rectangularQ[z], magnitudeRectangular[z],
           polarQ[z],       magnitudePolar[z],
           True,            Message[complex::badtype, z] ]

angle[z_] :=
    Which[ rectangularQ[z], angleRectangular[z],
           polarQ[z],       anglePolar[z],
           True,            Message[complex::badtype, z] ]
```

CmplxDis.m

number (either the real and imaginary parts or the magnitude and angle). The definitions are in
the program CmplxD1.m.

We read in the packages we need.	`In[1]:= << CmplxD1.m;\`
	`        << CmplxDis.m;\`
	`        << CmplxA1.m;`

Here is the number i in polar coordinates.	`In[2]:= i = makePolar[1, Pi/2]` `Out[2]= polar[1., 1.5708]`
Here is the number $1+i$ in Cartesian coordinates.	`In[3]:= j = makeRectangular[1, 1]` `Out[3]= rectangular[1., 1.]`
A sum is always returned in rectangular coordinates (because it is easier to compute that way).	`In[4]:= i + j` `Out[4]= rectangular[1., 2.]`
The product is expressed in polar coordinates.	`In[5]:= r = i j` `Out[5]= polar[1.41421, 2.35619]`
These are its magnitude and phase angle.	`In[6]:= {magnitude[r], angle[r]}` `Out[6]= {1.41421, 2.35619}`
These are its real and imaginary part.	`In[7]:= {realPart[r], imagPart[r]}` `Out[7]= {-1., 1.}`

```
(* rectangluar complex numbers *)

makeRectangular[x_, y_] := rectangular[N[x], N[y]]
rectangularQ[_rectangular] := True
rectangularQ[_] := False

realPartRectangular[rectangular[x_, y_]] := x
imagPartRectangular[rectangular[x_, y_]] := y
magnitudeRectangular[rectangular[x_, y_]] := Sqrt[x^2 + y^2]
angleRectangular[rectangular[x_, y_]] := ArcTan[x, y]

(* polar complex numbers *)

makePolar[r_, phi_] := polar[N[r], N[Mod[phi, 2Pi]]]
polarQ[_polar] := True
polarQ[_] := False

realPartPolar[polar[r_, phi_]] := r Cos[phi]
imagPartPolar[polar[r_, phi_]] := r Sin[phi]
magnitudePolar[polar[r_, phi_]] := r
anglePolar[polar[r_, phi_]] := phi

complexQ[z_] := rectangularQ[z] || polarQ[z]
```

CmplxD1.m

3.2.2　Data-Driven Programming

Instead of hardcoding a dispatch table into each polymorphic operation, we can put the names of the primitive operations into a global table, indexed by the name of the data type and the name of the (polymorphic) operation. Listing LookupTable.m shows the auxiliary data type for the table.

```
BeginPackage["LookupTable`"]

put::usage = "put[index1, index2, entry] inserts entry
    into the table."
get::usage = "get[index1, index2] returns the entry at the
    given position. If no entry is there, it returns Null."

Begin["`Private`"]

theTable[_, _] = Null     (* default *)

put[index1_, index2_, entry_] := theTable[index1, index2] = entry
get[index1_, index2_] := theTable[index1, index2]

End[]
EndPackage[]
```

LookupTable.m

The table is kept as a set of rules for the static variable `theTable`. This makes use of *Mathematica*'s built-in pattern matching. The default rule returns `Null` if there is no entry in the table for a given pair of indices.

The polymorphic operations are now defined in terms of an auxiliary procedure `operate` that accesses the table. This style of programming is called *data-driven programming*. The operations no longer contain conditional code, but are "driven" by the data that they receive as arguments. Finally, we have to fill in the table. The code is in CmplxTab.m.

```
(* Data-Driven Polymorphic Selectors *)

Needs["LookupTable`"]

complex::badtype = "no such operation or dataype: `1`, `2`."

operate[op_, data_] :=
    With[{basicop = get[Head[data], op]},
        If[basicop === Null,
            Message[complex::badtype, op, data],
            basicop[data] ]
    ]

(* polymorphic operations *)

realPart[z_] := operate[realPart, z]
imagPart[z_] := operate[imagPart, z]
magnitude[z_] := operate[magnitude, z]
angle[z_] := operate[angle, z]

(* fill in the table *)

put[ rectangular, realPart, realPartRectangular ]
put[ rectangular, imagPart, imagPartRectangular ]
put[ rectangular, magnitude, magnitudeRectangular ]
put[ rectangular, angle, angleRectangular ]

put[ polar, realPart, realPartPolar ]
put[ polar, imagPart, imagPartPolar ]
put[ polar, magnitude, magnitudePolar ]
put[ polar, angle, anglePolar ]
```

CmplxTab.m

We need these packages.	`In[1]:= << CmplxD1.m;\`
	`<< CmplxTab.m;\`
	`<< CmplxA1.m;`

Here is the number $1+i$ in Cartesian coordinates.	`In[2]:= j = makeRectangular[1, 1]` `Out[2]= rectangular[1., 1.]`
This is its fifth power.	`In[3]:= j^5` `Out[3]= polar[5.65685, 3.92699]`
Here is the real and imaginary part of the result.	`In[4]:= {realPart[%], imagPart[%]}` `Out[4]= {-4., -4.}`

3.2.3 Message Passing

So far, the "knowledge" of what data types and operations exist has either been spread out among the operations (with dispatch tables) or put into a global table (data-driven programming). A third method is to put the knowledge of the operations into the data objects themselves. Instead of calling a function `realPart` with a complex number object as the argument, we "call" the complex number object with the name of the operation (`realPart`) as the argument. The number object is then responsible for producing its real part. This style of programming is called *message passing*. If we implement our complex numbers this way, the implementation of the polymorphic operations becomes very easy. We keep the `operate` function from the preceding section, but define it so that it applies the data object to the operation:

$$\texttt{operate[op\_, data\_] := data[op]}$$

In a functional language such as LISP or *Mathematica* we can realize such objects by implementing them as pure functions, as we do in CmplxM1.m.

The constructors `makeRectangular` and `makePolar` return a pure function that takes one argument, `method`. Its body is a `Switch` statement that dispatches on the name of the method to return the appropriate result. Note that the branches of the switch were taken from the implementation of the primitive selectors in CmplxD1.m.

We read in the packages we need.

```
In[1]:= << CmplxM1.m;\
           << CmplxA1.m;
```

A complex number is now a complicated object.

```
In[2]:= i = makePolar[1, Pi/2]

Out[2]= Function[{method$},

                                      Pi                          Pi
           Switch[method$, realPart, 1 Cos[--], imagPart, 1 Sin[--],
                                      2                           2

                                  Pi
             magnitude, 1, angle, N[Mod[--, 2 Pi]], complexQ, True,
                                  2

             _, Message[complex::badmeth, method$]]]
```

All we can do with it is apply the known methods.

```
In[3]:= i[realPart]

Out[3]= 0
```

The arithmetic works just as before.

```
In[4]:= i^2 - i // Short

Out[4]//Short= Function[{method$}, <<1>>]
```

These are the real and imaginary parts of the previous result.

```
In[5]:= {realPart[%], imagPart[%]}

Out[5]= {-1., -1.}
```

Note that we have done away with the predicates `rectangularQ` and `polarQ`. There is no need to know the type of an object. Objects are completely defined by the methods we can apply to them; how they look internally is unimportant.

```
(* Complex Numbers by Message Passing *)

complex::badmeth = "unknown method `1`."

makeRectangular[x_, y_] :=
    Function[{method},
        Switch[method,
            realPart,   x,
            imagPart,   y,
            magnitude,  N[Sqrt[x^2 + y^2]],
            angle,      N[ArcTan[x, y]],
            complexQ,   True,
            _, Message[complex::badmeth, method]]
    ]

makePolar[r_, phi_] :=
    Function[{method},
        Switch[method,
            realPart,   r Cos[phi],
            imagPart,   r Sin[phi],
            magnitude,  r,
            angle,      N[Mod[phi, 2Pi]],
            complexQ,   True,
            _, Message[complex::badmeth, method]]
    ]

operate[op_, data_] := data[op]

(* polymorphic operations *)

realPart[z_]  := operate[realPart, z]
imagPart[z_]  := operate[imagPart, z]
magnitude[z_] := operate[magnitude, z]
angle[z_]     := operate[angle, z]
complexQ[z_]  := operate[complexQ, z]
```

CmplxM1.m

It may, at first, seem strange that we don't "see" the values of the complex numbers. In the previous two implementations we were presented with the internal forms of our objects. This is usually not a good idea. It can lead us into the temptation to access and modify objects directly instead of using selectors and constructors.

Our implementation is too low level. The pure functions used in many places should be hidden in a data type. Let us wrap them into an expression with head `complex`. A complex number is, therefore, represented by `complex[Function[{method}, ...]]`. The fact that complex numbers are now identified by the symbol `complex` also allows us to define a meaningful print form: we simply print them as *Mathematica* complex numbers. The improved code is in CmplxMsg.m. This approach also allows us to simplify the arithmetic definitions from CmplxA1.m. Since all complex numbers are now identified by the type `complex`, we can attach the rules to the symbol `complex` rather than to the arithmetic operators. The new version is in CmplxA.m.

```mathematica
(* Complex Numbers by Message Passing *)

complex::badmeth = "unknown method `1`."

makeRectangular[x_, y_] :=
    Function[{method},
        Switch[method,
            realPart,   x,
            imagPart,   y,
            magnitude,  N[Sqrt[x^2 + y^2]],
            angle,      N[ArcTan[x, y]],
            _, Message[complex::badmeth, method]]
    ]//complex

makePolar[r_, phi_] :=
    Function[{method},
        Switch[method,
            realPart,   r Cos[phi],
            imagPart,   r Sin[phi],
            magnitude,  r,
            angle,      N[Mod[phi, 2Pi]],
            _, Message[complex::badmeth, method]]
    ]//complex

operate[op_, complex[f_]] := f[op]

(* polymorphic operations *)

realPart[z_] := operate[realPart, z]
imagPart[z_] := operate[imagPart, z]
magnitude[z_] := operate[magnitude, z]
angle[z_] := operate[angle, z]

(* Output formatting *)

Format[c_complex] := realPart[c] + I imagPart[c]
```

CmplxMsg.m

We read in the packages we need.

```mathematica
In[1]:= << CmplxMsg.m;\
        << CmplxA.m;
```

We have defined our own print form for complex number objects. They print like ordinary complex numbers.

```mathematica
In[2]:= i = makePolar[1, Pi/2]

Out[2]= I
```

The output is deceptive since this is not the built-in complex number I:

```mathematica
In[3]:= InputForm[%]

Out[3]//InputForm=
    complex[Function[{method$},
        Switch[method$, realPart, 1*Cos[Pi/2], imagPart,
            1*Sin[Pi/2], magnitude, 1, angle, N[Mod[Pi/2, 2*Pi]],
            _, Message[complex::badmeth, method$]]]]
```

The arithmetic works as always.

```mathematica
In[4]:= i^2 - i

Out[4]= -1. - 1. I
```

```
(* Polymorphic Complex Arithmetic *)

complex/: z1_complex + z2_complex :=
    makeRectangular[ realPart[z1] + realPart[z2],
               imagPart[z1] + imagPart[z2] ]

complex/: -z_complex := makeRectangular[ -realPart[z], -imagPart[z] ]

complex/: z1_complex * z2_complex :=
    makePolar[ magnitude[z1] * magnitude[z2],
               angle[z1] + angle[z2] ]

complex/: (z_complex)^n_Integer :=
    makePolar[ magnitude[z]^n, n angle[z] ]
```

CmplxA.m

If we anticipate that calls to selectors are much more frequent than constructor calls, we could precompute certain values and thus speed up the selectors. For rectangular numbers, computing the magnitude and angle are time consuming. We could, therefore, change the constructor like this:

```
makeRectangular[x_, y_] :=
    With[{mag = N[Sqrt[x^2 + y^2]], ang = N[ArcTan[x, y]]},
        Function[{method},
            Switch[method,
              realPart,  x,
              imagPart,  y,
              magnitude, mag,
              angle,     ang,
              _,          Message[complex::badmeth, method]]
        ]
    ]//complex
```

An alternative for <code>makeRectangular</code>

3.2.4 Comparison of the Different Implementations

We have seen three different implementations of polymorphic data types, using dispatch tables, data-driven programming, and message passing. Each has, of course, the same interface to the outside world, namely the polymorphic selectors realPart, imagPart, magnitude, and angle. To compare the relative merits and disadvantages of the three methods we mention briefly several questions.

Where are decisions taken?
   With dispatch tables, each operation dispatches on the type of the operands (which must therefore be known). The data objects are passive. With message passing, the dispatch

happens inside the active objects. The operations are passive. Data-driven programming is somewhere in the middle. The interface between data and operations is in one global table.

How can we add new data types?
To add more data types to a system using dispatch tables, we have to go through each operation and add another branch to the Which statement inside. This is clearly not a very modular design. In a message passing system we simply define the constructor for the new data type and put in implementations for all methods. None of the other code needs to be touched. In a data-driven system we add a new *row* to the dispatch table and define the necessary functions.

How can we add new operations?
The roles of data and objects are now reversed. It is easy to add new operations to a dispatch table system, but cumbersome to do so in a message passing system. In a data-driven system we add a new *column* to the dispatch table.

The implementation to choose depends therefore on its anticipated evolution. Do we expect to add more data types or rather new functions on the existing types? It is relatively rare to add new operations to all existing types. Usually new types come with their own new functions. Furthermore, there is an elegant way to group common operations together for several similar data types (the method complexQ in CmplxMsg.m, for example, is common to both rectangular and polar complex numbers). This idea leads to the area of *object-oriented programming*, which will be the topic of Chapter 4. In the remainder of this chapter we will look at some more issues regarding message passing and discuss data-driven programming in *Mathematica*'s rule-based environment.

3.3 Data-Driven Programming and Rules

The implementations in the preceding section have a LISP-like flavor. Indeed the example has been inspired by a similar treatment of the material in the LISP dialect SCHEME, taken from the book *Structure and Interpretation of Computer Programs* [1].
The Which statements inside the polymorphic selectors realPart, etc., in CmplxDis.m are not good *Mathematica* programming style. Much better is a set of rules such as the following:

```
realPart[z_?rectangularQ] := realPartRectangular[z]
realPart[z_?polarQ] := realPartPolar[z]
```

Rule-based dispatch tables (excerpt)

Similar rules can be given for the other functions imagPart, magnitude, and angle.
This style of programming blurs the distinction between polymorphism and overloading. Our first implementation of realPart was truly polymorphic, taking arguments of different types.

Writing it as a set of separate rules is more like *overloading,* where we provide several definitions for *one* function symbol. *Mathematica* resolves the overloading by pattern matching. Compiled languages like C++ resolve it by looking at the type of the argument. (These languages do not allow typeless arguments as *Mathematica* does.)

Knowing that the rectangular and polar complex numbers are implemented as a simple data type identified by the heads `rectangular` and `polar` (see CmplxD1.m), we can simplify the rules a bit:

```
realPart[z_rectangular] := realPartRectangular[z]
realPart[z_polar] := realPartPolar[z]
```

Faster rule-based dispatch tables (excerpt)

These rules are stored with the symbol `realPart`. What we have done, therefore, is to store the *columns* of the lookup table with their respective symbols (compare with CmplxTab.m).

As an alternative, we can easily store the rules with the names of the *data types:*

```
rectangular/: realPart[z_rectangular]  := realPartRectangular[z]
rectangular/: imagPart[z_rectangular]  := imagPartRectangular[z]
rectangular/: magnitude[z_rectangular] := magnitudeRectangular[z]
rectangular/: angle[z_rectangular]     := angleRectangular[z]
```

Rule-based message passing (excerpt)

The rules for the type `polar` are similar. This approach corresponds to storing the *rows* of the lookup table with their respective symbols. It is simply a variant of the message passing style, where we store all functions with the data types instead of with the operations.

Which of the two ways is better? In the first case (using downvalues) the rule list for one operation contains an entry for each data type. Searching these rules will get slower as the number of data types increases. In the second case (using upvalues) there is one rule list for each data type and it grows with the number of operations. As an example, the built-in arithmetic operations take potentially many data types, so it is usually better to store these rules with the data type.

3.4   A Better Interface for Objects

We can get rid of definitions such as `realPart[z_] := operate[realPart, z]` for each method in CmplxMsg.m by defining a general rule for applying methods to objects. Instead of putting the dispatch function inside the object, the constructor returns only a symbol and defines the necessary rules for it. The new code for rectangular complex numbers is in Rect-Objects.m. We use `Module[{symbol}, body]` to create a unique symbol for each call of

```
makeRectangular[x_, y_] :=
   Module[{rectangular},
    With[{dispatch = Function[ {method},
                       Switch[method,
                          realPart,   x,
                          imagPart,   y,
                          magnitude, N[Sqrt[x^2 + y^2]],
                          angle,      N[ArcTan[x, y]]
                       ] ]},
        rectangular/: f_Symbol[rectangular] := dispatch[f] /;
           MemberQ[{realPart, imagPart, magnitude, angle}, f];
        rectangular
    ]]
```

RectObjects.m

`makeRectangular`. The rule for applying methods to this object is then stored with that symbol. The inner `With` causes the body of the dispatch function to be textually inserted into the rule. Note that `Module` is normally used to generate local program variables. These variables are removed after leaving the scope of the module because they are no longer referenced. Here we do not give the local variable `rectangular` a value. Since we return it as the value of the body of the module, it is exported outside of its defining module and is not removed. It gives no information about the object and can only be passed around and used for invoking methods. Such a value is sometimes called a "cookie" (unlike real ones, you cannot eat these).

Let us look at some examples.

```
In[1]:= << RectObjects.m
```

`makeRectangular` returns just a symbol.

```
In[2]:= i = makeRectangular[1, 1]
Out[2]= rectangular$1
```

You can see the rule attached to it in this way.

```
In[3]:= ?rectangular$1
Global`rectangular$1

Attributes[rectangular$1] = {Temporary}

(f$_Symbol)[rectangular$1] ^:=
  Function[{method$}, Switch[method$, realPart, 1,
      imagPart, 1, magnitude, N[Sqrt[1^2 + 1^2]], angle,
      N[ArcTan[1, 1]]]][f$] /;
   MemberQ[{realPart, imagPart, magnitude, angle}, f$]
```

It responds to the methods we defined.

```
In[3]:= realPart[i]
Out[3]= 1

In[4]:= magnitude[i]
Out[4]= 1.41421
```

It does nothing with other functions.

```
In[5]:= f[i]
Out[5]= f[rectangular$1]
```

3.5   Objects with Local State

The objects we have looked at so far have been very simple. The results from all methods could be computed from the values given to the constructor. In constructing a `rectangular` object, *Mathematica*'s mechanism for rule definitions inserts the values of the pattern variables x and y into the body of the rule. (Recall that pattern variables are not local variables in the sense of Pascal or **C**.) Furthermore, all our methods were parameterless procedures. Things get more complicated if we allow objects to have an internal state that can be modified by certain methods.

An example is a savings account. It is opened with an initial balance by `makeAccount[`*bal*`]`. There are methods to deposit and withdraw amounts from it. These methods change the internal state (the balance). We cannot insert the value of the initial balance textually into the body of the dispatch routine. We need to keep a symbol around to which we bind the current balance value. The methods `deposit[`*account, amount*`]` and `withdraw[`*account, amount*`]` each have one parameter (besides the account object, of course). The file AccountObjects.m contains a solution to these problems.

```
makeAccount[bal0_] :=
    Module[{account, bal = bal0},
     With[{dispatch = Function[ {method},
                       Switch[method,
                          balance,   bal&,
                          deposit,   Function[{am}, bal += am],
                          withdraw,  Function[{am}, bal -= am]
                       ]
                  ]},
         account/: f_Symbol[account, args___] :=
                     dispatch[f][args] /;
                  MemberQ[{balance, deposit, withdraw}, f];
         account
     ]
    ]
```

AccountObjects.m

We declare one more variable `bal` in the `Module` and initialize it with the opening balance. The values of all methods are themselves functions, which are then applied to the arguments of the method call (if any). The balance enquiry `balance` has no arguments and is therefore implemented as a constant pure function.

Time for an example!

```
In[1]:= << AccountObjects.m
```

We create a new account with an opening balance of $100.

```
In[2]:= a = makeAccount[100]

Out[2]= account$1
```

Now we deposit $50 into this account. The new balance is returned.

```
In[3]:= deposit[a, 50]
Out[3]= 150
```

Withdrawing an amount decreases the balance, as we all know very well.

```
In[4]:= withdraw[a, 99]
Out[4]= 51
```

The parameterless method `balance` simply returns the current value of the balance variable.

```
In[5]:= balance[a]
Out[5]= 51
```

Note that the balance is not kept in the body of the switch statement (there is only a reference to the variable bal$1 there).

```
In[6]:= ?account$1

Global`account$1

Attributes[account$1] = {Temporary}

(f$_Symbol)[account$1, args$___] ^:=
  Function[{method$}, Switch[method$, balance, bal$1 & ,
      deposit, Function[{am$}, bal$1 += am$], withdraw,
      Function[{am$}, bal$1 -= am$]]][f$][args$] /;
    MemberQ[{balance, deposit, withdraw}, f$]
```

The balance is kept as the value of bal$1.

```
In[6]:= ?bal$1

Global`bal$1

Attributes[bal$1] = {Temporary}

bal$1 = 51
```

Another account has a distinct balance variable.

```
In[6]:= b = makeAccount[1000]
Out[6]= account$4
```

The balance of a is not disturbed.

```
In[7]:= balance[a]
Out[7]= 51
```

Our examples use the fact that `Module` creates new variables each time it is executed. We can see these variables if we return them from a module without giving them a value.

```
In[8]:= Module[{var}, var]
Out[8]= var$5
```

The counter `$ModuleNumber` is used to create unique new variables of the form var$$n$. Each time a module is executed, the counter is incremented.

```
In[9]:= $ModuleNumber
Out[9]= 6
```

In certain dialects of LISP (SCHEME being one of them) the normal binding method for *lambda* variables keeps different bindings in different invocations of a function separate. The penalty for this is usually a slower variable lookup. In *Mathematica* and some versions of LISP you can decide for yourself whether you want the faster substitution semantics (as in the RectObjects example) or whether you need static scoping as in the preceding account example.

3.6   Conclusions

We have discussed different ways of implementing polymorphic operations. Such operations can be realized in *Mathematica* very elegantly. The main idea is to use upvalues to give new rules for existing functions operating on new data types. We also looked at objects, that is, data elements that contain embedded knowledge about the operations that can be performed on them. We will extend these ideas in a discussion of object-oriented programming in the next chapter.

 The floppy disk contains the packages CmplxA1.m, CmplxDis.m, CmplxD1.m, Lookup-Table.m, CmplxTab.m, CmplxM1.m, CmplxMsg.m, CmplxA.m, RectObjects.m, and AccountObjects.m.

Chapter Four
Object-Oriented Programming

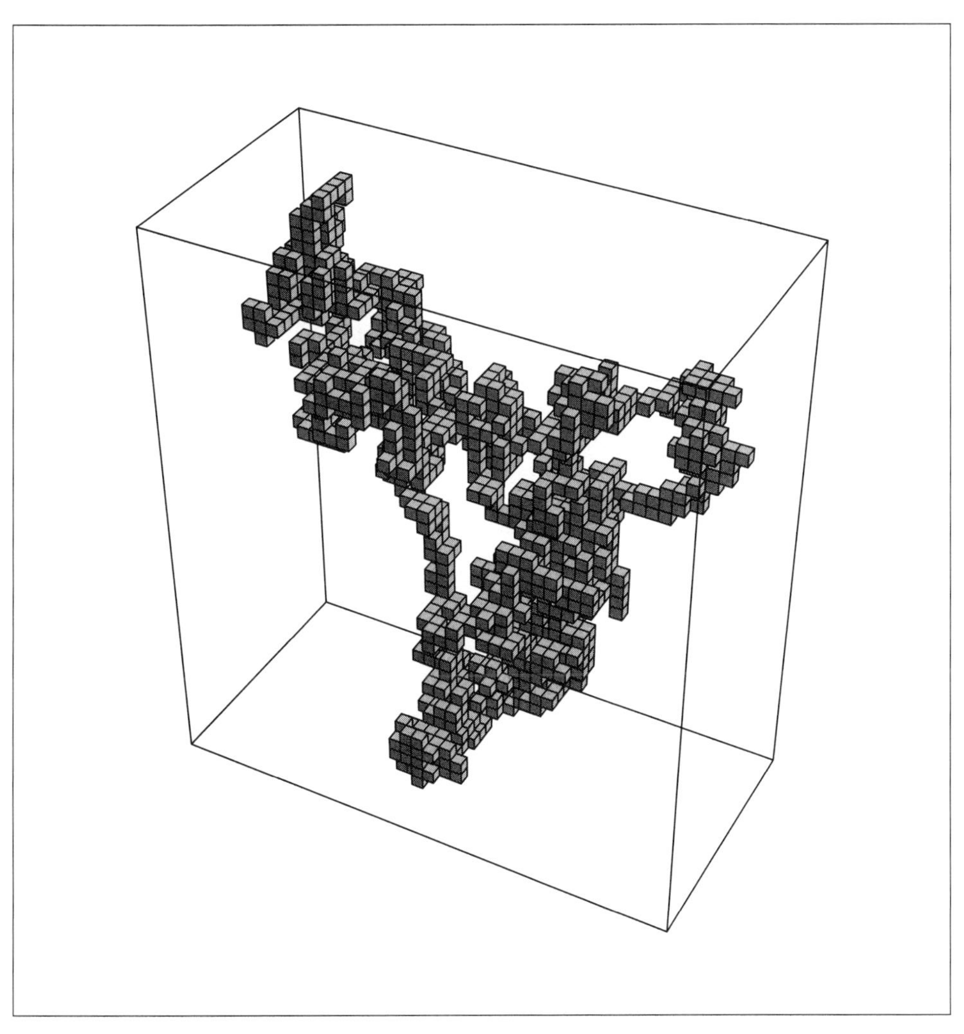

The object-oriented programming style is becoming increasingly popular. It promises code reuse and easier maintenance of larger projects than is possible with traditional procedural languages. Its use of methods and message passing instead of procedure calls shifts the programmer's view toward close integration of data and operations. An interactive object-oriented language can easily be implemented in *Mathematica*. In the last chapter we discussed two important tools, message passing and objects. We now present the other tools, classes and inheritance. To illustrate the programming style we expand the example developed in the last chapter and give an implementation of *collections*.

About the illustration overleaf:
A random walk in three dimensions. At each stage, a random direction among the six orthogonal directions is chosen. The path is shown by placing unit cubes in the locations visited. The simple code for this example is in the package RandomWalk3D.m.

4.1   The Object-Oriented Paradigm

One step toward better organization of large programming projects was *modularization,* whereby related functions and data definitions are grouped together and provided with a defined interface to the rest of the code. Implementation and interface are thus separated. Ideally, users of a module would not need to know how it is implemented, only how to use it. In *Mathematica* modularization is implemented with the context-manipulation mechanisms `BeginPackage` and `EndPackage` in a *package.* (`Module` in *Mathematica* provides encapsulation on the smaller scale of an individual procedure.)

Higher level data types (especially *records* or *structs*) allow one to treat collections of data as a unit. Abstract data types provide a clean foundation for this idea. We have shown how they can be used in *Mathematica* in Chapter 2. At this point we observe an asymmetric treatment of data and functions. In most languages, functions cannot be members of data types. The task of deciding which function to apply to a data object has to be encoded within the body of the function. As a consequence the code is sprinkled with conditional statements, dispatch tables, etc.

The first important aspect of object-oriented languages is that functions are considered part of data. A data object "knows" which operations can be performed on it. The functions defined for a certain type of object are part of that object. Thus an *object* is a collection of data elements and operations that act on these data elements. The operations are called *methods.* A uniform mechanism, called *message passing,* is provided for invoking the correct piece of code when a function is called, or as we now prefer to say, a message is passed to an object to execute a certain method. Methods are usually not defined for each object separately but are collected in a *class.* Objects then belong to a class from which they take their methods.

The second important aspect is *inheritance.* Often a number of related data types have some common characteristics. Some operations on them can be written in a way that does not depend on which of the data types they are applied to. Common characteristics of related data types can then be isolated and encapsulated in a new data type. The related data types are made subtypes of the new type. They inherit the characteristics of the common type and need only implement those aspects in which they differ from their supertype. Thus much of the code needs to be written only once. This saves development time and—perhaps more important—ensures consistency, since a change needs to be made only once, instead of being applied to several almost identical pieces of code.

4.1.1   Message Passing and Methods

In Chapter 3 we looked at message passing as an implementation method for polymorphism. Let us now look at message passing in a more general sense, as a way of organizing data and operations coherently. Most operations to be performed on data have among their arguments one particular object that they modify or inspect. We can view this as a request to the object to perform a certain operation, possibly modifying its internal state. The notion of "internal state" implies that these internal details should be encapsulated and accessible only through a defined interface. This is again modularization, now on the level of individual data objects. The methods provide the interface. Budd defines messages and methods in *An Introduction to Object-Oriented Programming* [6] as follows:

> Action is initiated in object-oriented programming by the transmission of a *message* to an agent (an *object*) responsible for the action. The message encodes the request for an action, and is accompanied by any additional information (arguments) needed to carry out the request. . . . In response to a message, the receiver will perform some *method* to satisfy the request.

In functional languages there is one particularly nice implementation of objects. The data objects are themselves *functions* that receive the name of the method and any additional parameters as arguments and then perform the code belonging to the method. At the end of the last chapter we presented an example defining *account objects*. Accounts.m is a slightly different implementation that is more suitable for the extensions we will need later.

```
accountvars = {bal}  (* instance variables *)
accountmethods = {   (* methods *)
    {new,      Function[init, bal = init]},
    {balance,  bal&},
    {deposit,  Function[am, bal += am]},
    {withdraw, Function[am, bal -= am]}
}

(* names of methods == messages *)
messages = First /@ accountmethods
Apply[ (switch[#1] = #2)&, accountmethods, {1} ]

apply[meth_, obj_, args___] :=
    With[{lvars = variables[obj], ivars = Hold @@ accountvars},
        (switch[meth] /. List @@ Thread[ ivars :> lvars, Hold]
        )[args]
    ]

makeAccount[bal0_] :=  (* the constructor *)
    Module[{account},
        account/: variables[account] =
                Hold @@ Unique[Evaluate[accountvars]];
        account/: f_Symbol[account, args___] :=
                apply[f, account, args] /; MemberQ[messages, f];
        new[ account, bal0 ];  (* initialize it *)
        account
    ]
```

The internal state of an object is defined by a list of variables, called *instance variables,* that should be local to each object. Here we have only one such variable, the balance `bal`. The methods are given as a list of pairs, the first element of each pair is the name of the method, the second element is the (pure) function that implements the method. Methods that simply inspect the internal state, for example returning the balance, are written as functions without any additional arguments. The method `new` is automatically called by the constructor `makeAccount` to initialize the internal state of each new object. The statement

```
Apply[(switch[#1] = #2)&, accountmethods, {1}]
```

defines the methods as values of the symbol `switch` in such a way that `switch[`*meth*`]` gives the function used to implement the method named *meth.*

Let's see how this works.

```
In[1]:= accountmethods = {
            {new,      Function[init, bal = init]},
            {balance,  bal&},
            {deposit,  Function[am, bal += am]},
            {withdraw, Function[am, bal -= am]}
        };
```

Here we define the values for the symbol `switch` in the form `switch[`*meth*`] = Function[...]`.

```
In[2]:= Apply[ (switch[#1] = #2)&, accountmethods, {1} ];
```

You can see the resulting rules for `switch`.

```
In[3]:= ?switch

Global`switch

switch[balance] = bal &

switch[deposit] = Function[am, bal += am]

switch[new] = Function[init, bal = init]

switch[withdraw] = Function[am, bal -= am]
```

The auxiliary function `apply` performs the message passing. It receives the name of the method, the object, and any additional parameters as arguments. First, we find the function belonging to the method, as described earlier (using `switch[meth]`). Each object will have its own symbols for the instance variables. Since the method was defined using the generic symbols from the list `accountvars`, we have to substitute them before evaluating the method body. For this, we build a list of rules to substitute the particular symbols used for the instance variables of the given object (in `lvars`) for the generic variables (in `ivars`). Both variable "lists" use `Hold` as their head to avoid evaluation. The construct `Thread[ ivars :> lvars, Hold]` transforms the rule of these "holds" into a "hold" of rules. Since we use delayed rules, the protection offered by `Hold` is no longer needed so we replace this head with `List`. Using this list of rules, we perform the replacement inside the body of the function. Finally, the function is applied to any additional parameters given.

The constructor `makeAccount` creates a new account object. An object is simply a new symbol, created implicitly inside `Module`. The constructor then creates new unique symbols for the instance variables and stores their names as the value of `variables[`*object*`]`. Next, the rule for applying functions to the object is defined. It uses `apply` to pass the function name as a message to the object. Finally, the method `new` is invoked to initialize the object.

We create and initialize an object with the function `makeAccount`.

```
In[3]:= a = makeAccount[100]

Out[3]= account$1
```

Here we see the two rules defined for the object. The rule for `variables` defines the local symbols for this particular account. The message passing mechanism is implemented as a rule for applying functions to objects.

```
In[4]:= ??account$1

Global`account$1

Attributes[account$1] = {Temporary}

variables[account$1] ^= Hold[bal$2]

(f$_Symbol)[account$1, args$___] ^:=
    apply[f$, account$1, args$] /; MemberQ[messages, f$]
```

The message passing mechanism is implemented as a rule for applying functions to objects. A method is called by writing it as an ordinary function call with the object as its first argument. The rule turns this into a call of `apply` if the function is one of the known methods.

We can increase the balance of the account by depositing some money into it.

```
In[4]:= deposit[a, 50]

Out[4]= 150
```

If we use a function that is not one of the known methods a normal evaluation takes place. Here, we use the built-in function `Attributes` to get the list of attributes of the symbol `account$1`.

```
In[5]:= Attributes[account$1]

Out[5]= {Temporary}
```

4.1.2 Inheritance

Different data types are usually not completely independent. Often, certain objects can be viewed as specializations (subtypes) of others. An example with deep and richly structured subtype relationships is a modern *window system* (NeXTstep, OpenWindows, Motif). The objects on the screen are all windows of different sorts. There are menus, attention panels, and "windows" proper. Windows consist of different parts: title bars, window border, scroll bars, button areas, and panes (subwindows). Panes are of different types: text panes, PostScript viewing areas, custom (application-specific) panes. A window system is *event driven*. Each user action (moving the mouse, pressing down a button, hitting a key) is an event. The *event loop* determines which object on the screen should respond to an event. Message passing is a natural way of implementing such a system. The event loop doesn't need to know the type of the object that receives the event. It simply sends a message "mouse key down" to the frontmost window under the cursor. That window in turn determines which of its parts lies under the cursor and passes the message on to this object. Eventually the message arrives at a button object, for example. The button then changes its visual representation on the screen and invokes a particular method in the applications program. (When programming the NeXTstep system with the *Interface Builder*, one can even make such connections graphically; the code is generated automatically.)

In an object-oriented approach, a feature that is common to several objects is isolated in an *abstract class* and programmed only once. All different kinds of objects that use that feature are made subclasses of this abstract class. In our window system we would have an abstract class window, of which the different kinds of windows are subclasses, for example.

A *class* is a definition of a type consisting of instance variables and methods. The type account in the preceding section is a class. A class provides at least one "factory method," a method for creating *instances* of that class. makeAccount is the factory method in our previous example; it returns a new object of type account. An *instance* of a class is an object. It has its own set of the instance variables. When a message is sent to an object it performs one of the methods defined in its class. In our example we defined methods named balance, deposit, and withdraw. There is also a special method called new, which is called by the constructor to initialize the objects.

Each class also has a *superclass* as one of its attributes. All instance variables and methods of the superclass (and its superclass, and so on) are considered part of the instance variables and methods of the instances of a class. They are called *inherited instance variables* and *inherited methods*. When a message is sent to an object and no method is defined for it in the class of the object, it is *passed along* to the superclass. Eventually the message will reach a class where it is defined and the code of its method is executed.

A *class method* is a method defined for classes themselves, not for objects. The class method SuperClass[*class*], for example, is commonly defined and returns the superclass of a given class. Some languages also provide *class variables*. These are variables whose values are shared by all objects of the class. They exist only once per class, unlike the instance variables.

A subclass can *override* a method of its superclass by redefining it. If a message is sent to this subclass its own method is invoked, not the one in the superclass. Many object-oriented programming languages offer a way to invoke a method from the superclass directly, bypassing this scheme. This is usually done by providing a pseudovariable super that denotes the receiver of the currently performed method (the object on which the method operates) viewed as a member of its superclass. Another pseudovariable, usually called self or this, can be used inside the code of a method to send messages to the receiver of the message currently being performed.

We have said that a message not defined in a class can be passed along to the superclass. Eventually there must be an end to the chain of classes. The universal class of which every other class is a direct or indirect subclass is usually called Object; it provides default implementations of the standard methods.

4.2 An Implementation in *Mathematica*

We can build the primitive class definition shown in the `account` example into a general mechanism for defining classes with inheritance. The procedure for applying methods becomes more complicated, since it has to implement passing messages to the superclass. We also need to treat the variables `self` and `super`.

New classes will be declared with the procedure

$$\texttt{Class}[\textit{class, superclass, variables, methods}]\,,$$

where *class* is the name of the new class being defined, *superclass* is its superclass, *variables* is the list of instance variables, and *methods* is the list of methods in the form *name*, *body*.

We will define the following class methods:

`SuperClass[`*class*`]`	the superclass of *class*
`Methods[`*class*`]`	all methods known by *class*
`InstanceVariables[`*class*`]`	all instance variables of *class*
`new[`*class*`, args...]`	create an instance of *class*
	(the arguments are used for initialization)

The following standard methods are defined for all objects of a class:

`Class[`*obj*`]`	the class of *obj*
`SuperClass[`*obj*`]`	the superclass of the class of *obj*
`Methods[`*obj*`]`	the methods of the class of *obj*
`InstanceVariables[`*obj*`]`	the instance variables of *obj*
`isa[`*obj*`, class]`	true, if *obj* is of the given class
	(or one of its subclasses)
`new[`*obj*`, args...]`	initialize *obj*
	(usually called by the factory method `new` above)

The listing at the end of this chapter shows the rather sophisticated code of the package Classes.m, which provides a full object-oriented system for *Mathematica*. If you are not interested in the details of the implementation, you may want to skip directly to the next section, which gives a simple example of how it works.

Note that this implementation is substantially different from the first version published in the *Mathematica Journal* [30].

The main procedure defined in Classes.m is `Class[`*class*`, `*superclass*`, `*variables*`, `*methods*`]`, used to declare new classes. It defines the factory method `new` for *class* and the other class methods `Methods`, `InstanceVariables`, and `SuperClass`. It also precomputes as many values for the new class as possible. Therefore, we also choose a different representation of objects. Instead of using a symbol, we use an expression of the form *cookie*[*var$_1$*, ...], where the *var$_i$* are the symbols used for the instance variables (as before). *cookie* is a symbol derived from the name of the class. It is the same for all objects.

The procedure starts out by adding two standard methods, `Class` and `isa`, to the methods given for the new class. The class methods are defined as simple rules. An auxiliary predicate, `methodQ`, is defined for later use. The auxiliary function `apply` implements message passing. One rule is defined for it for each local method (similar to `switch` before). Another rule passes unknown messages to the superclass. All of these definitions are inside a `With` construct that serves to insert some values into the otherwise unevaluated bodies of the rules.

The main body for the rule for applying local methods is much like the corresponding function in our accounts example. The only difference is the treatment of `self`, which is simply replaced by the object, and `super`, which bypasses the normal method lookup and calls the apply procedure of the superclass directly.

The auxiliary class method `methodHandler` makes the function `apply` available for the use in subclasses. You can see how it is used in the inheritance rule for `apply`. The rule for message passing is now defined for the symbol `cookie`. There is therefore only one such rule per class, and no rules need to be given for the objects themselves. There is also a rule for implementing `super`. It is attached to the auxiliary symbol `raise`.

Next, we define the factory method `new`. `Unique[]` is used to create the symbols that hold the values of the instance variables of the new object. These variables are given the attribute `Temporary`, which will cause them to be destroyed whenever the new object is no longer in use. The new object is then initialized by calling the method `new`.

Finally, we define a print form for objects that hides their internals.

To define the class `Object`, we set up a dummy superclass that contains the essential minimum set of class methods and then call the procedure `Class` for the first time to create `Object`. Observe that among the methods for `Object` are many of the standard methods that each class has. We use the new mechanism of inheritance here, too.

An interesting detail of the implementation is the following construct:

```
With[{allvariables = allvariables, ...},
   :
   :
];
```

The expression `With[{var = var}, code]` simply inserts the current value of *var* everywhere inside *code*, even into the bodies of definitions that are not evaluated. You may want to think about how this works.

4.2.1   Accounts as Classes

Let us now show the account example in the framework of object-oriented programming. The following class definitions can be found in the package AccountClasses.m.

Here is the account class again, now defined as a subclass of `Object`. The method `new` first calls the same method of its superclass. This is recommended to make sure that the object is properly initialized, even if we know (by peeking at the implementation) that the default implementation for `new` in class `Object` does nothing.

```
In[1]:= Class[ Account, Object,
             {bal},
             {{new,        (new[super]; bal = #1)&},
              {balance,    bal&},
              {deposit,    Function[bal += #1]},
              {withdraw,   Function[bal -= #1]}
             }
        ];
```

An account is created by a call to the factory method `new`.

```
In[2]:= a1 = new[Account, 100]

Out[2]= -Account-
```

Message passing looks like an ordinary function call. The method `balance` is invoked and returns the value of the instance variable `bal`.

```
In[3]:= balance[a1]

Out[3]= 100
```

We now declare a subclass of `Account` to implement accounts that may not have a negative balance. We simply override the method `withdraw`. It first checks whether the balance is sufficient for the withdrawal. If it is, we call the method `withdraw` in the superclass, otherwise we print an error message. All other behavior of this class is inherited.

The pseudo-object `super` stands for `self`, but temporarily viewed as a member of its superclass. Therefore, the ordinary code for withdrawals in the class `Account` takes effect.

```
In[4]:= Class[ NoCreditAccount, Account,
             {},
             {{withdraw,
                   Function[
                       If[ balance[self] < #1,
                           Print["cannot withdraw ", #1],
                           withdraw[super, #1] ] ]}
             }
        ];

In[5]:= b1 = new[NoCreditAccount, 1000]

Out[5]= -NoCreditAccount-
```

If the balance is sufficient, withdrawal works as always.

```
In[6]:= withdraw[b1, 900]

Out[6]= 100
```

If the balance is insufficient, we get an error message.

```
In[7]:= withdraw[b1, 110]

cannot withdraw 110
```

All other methods are inherited from the class `Account`.

```
In[8]:= balance[b1]

Out[8]= 100
```

Here is another type of account, one that requires a minimum balance. It is created with `new[MinimumBalanceAccount, `*balance*`, `*minimum*`]`.

Note how we pass the first argument for initialization on to the method new in the superclass. The second initialization argument is used for the additional instance variable min.

```
In[9]:= Class[ MinimumBalanceAccount, Account,
              {min},
              {{new,        (new[super, #1]; min = #2)&},
               {withdraw,
                   Function[
                       If[ balance[self] - min < #1,
                           Print["cannot withdraw ", #1],
                           withdraw[super, #1] ] ]}
              }
       ];
```

Two arguments need to be supplied. The second one initializes the extra instance variable min.

```
In[10]:= c1 = new[MinimumBalanceAccount, 1000, 100];
```

Again, the balance is checked and if insufficient, the withdrawal is not performed.

```
In[11]:= withdraw[c1, 901]
```

```
cannot withdraw 901
```

4.3   Abstract Classes and Polymorphic Data

The use of abstract classes is an elegant solution to the problem of polymorphic data types. Here, then, is an object-oriented version of the polymorphic complex numbers from Section 3.1. As a reminder: We want to represent complex numbers either in Cartesian or in polar coordinates, but use a common interface. All the code shown in this section is in CmplxCl.m.

The common interface is put into an abstract class complex:

```
Class[ complex, Object,
     {},
     { {realPart,    (magnitude[self] Cos[angle[self]])&},
       {imagPart,    (magnitude[self] Sin[angle[self]])&},
       {magnitude,   (Sqrt[realPart[self]^2 + imagPart[self]^2])&},
       {angle,       (ArcTan[realPart[self], imagPart[self]])&}
     }
]
```

The abstract class complex

The class declares the (polymorphic) selectors as methods and provides the default implementation, which expresses the real and imaginary parts in terms of magnitude an angle, and vice versa. This is, of course, a circular definition. The circle will be broken by specialization. Note that there is no constructor new[]. This is characteristic for abstract classes: There will never be any object belonging to this class.

The Cartesian complex numbers are now declared as a subclass of complex. They have two instance variables, the real and imaginary parts:

```
Class[ cartesian, complex,
     {x, y},
     { {realPart,        x&},
       {imagPart,        y&},
       {new,             ({x, y} = {##})&}
     }
]
```

Cartesian complex numbers

Since the real and imaginary parts are stored as instance variables, we can override the default implementation of `realPart` and `imagPart` to simply return the values of these instance variables. The constructor initializes the instance variables in the usual way.

An analogous class definition is made for the polar complex numbers. Here we override the default implementation of `magnitude` and `angle`.

```
Class[ polar, complex,
     {r, phi},
     { {magnitude,       r&},
       {angle,           phi&},
       {new,             ({r, phi} = {##})&}
     }
]
```

Polar complex numbers

Here is the number $1+i$ in Cartesian coordinates.	`In[1]:= z = new[cartesian, 1, 1]` `Out[1]= -cartesian-`
Its real part is found using the method in the class `cartesian`.	`In[2]:= realPart[z]` `Out[2]= 1`
The magnitude is computed with the default method in class `complex`.	`In[3]:= magnitude[z]` `Out[3]= Sqrt[2]`

Arithmetic can now be defined in the same way as in CmplxA1.m (see page 38). The predicate `complexQ[ ]` simply tests membership of the class `complex`. The code is shown at the top of the following page.

The square is computed in polar coordinates.	`In[4]:= z^2` `Out[4]= -polar-`
Here are its real and imaginary parts.	`In[5]:= {realPart[%], imagPart[%]}` `Out[5]= {0, 2}`

```
complexQ[z_] := isa[z, complex]

unprotected = Unprotect[Plus, Times, Power]

z1_?complexQ + z2_?complexQ :=
    new[ cartesian, realPart[z1] + realPart[z2], imagPart[z1] + imagPart[z2] ]

-z_?complexQ := new[ cartesian, -realPart[z], -imagPart[z] ]

z1_?complexQ * z2_?complexQ :=
    new[ polar, magnitude[z1] * magnitude[z2], angle[z1] + angle[z2] ]

(z_?complexQ)^n_Integer := new[ polar, magnitude[z]^n, n angle[z] ]

Protect[ Evaluate[unprotected] ];
```

Arithmetic with complex numbers

4.4 An Advanced Example: Collections from SMALLTALK

Collections are data types that are used to hold other data elements in various ways. Examples are lists and sets. *Indexed collections* allow the retrieval of their elements through a key value. Examples are arrays, where the keys are integers, and dictionaries, where the keys are arbitrary values. The object-oriented language SMALLTALK offers an extensive set of collections. Functions defined for collections include methods to test for membership, to add or delete elements, a general iterator construct that applies a function to all elements in a collection in turn (like Map in *Mathematica*), and a reduction operation similar to Fold.

In this section we discuss an implementation of SMALLTALK collections in *Mathematica*. The version of SMALLTALK collections we use is the one presented in Budd's book [6]. The class hierarchy has the structure depicted in Figure 4.4–1. collection and indexedCollection are abstract base classes. Table 4.4–1 shows all methods defined for collections. The code for all collections described here is in the package Collections.m.

Remarkably, all of the first group of methods can be implemented in terms of do. Only do needs to be written in each subclass of collection. The other methods all have a default implementation in the abstract class collection. Listing collection shows the definition of this class.

The method fold uses a local variable to hold the value produced by *binop* and iterates over the collection. includes searches through the collection for an occurrence of the element and, if it finds it, immediately returns since there is no point in continuing the search; this is done with a nonlocal return using Throw and Catch. The size is computed by folding the function that adds 1 to its first argument. The collection is empty if its size is 0. To select elements we build up a list with the elements that satisfy the given predicate.

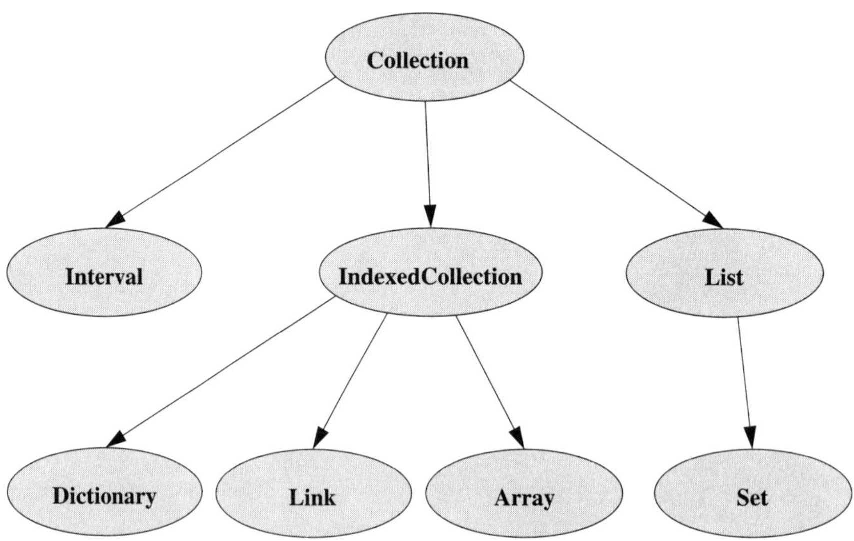

Figure 4.4–1: The class hierarchy of collections

These are the methods that are available for collections:

do[*coll*, *unop*]	apply the unary function *unop* to all elements
size[*coll*]	return the number of elements in *coll*
isEmpty[*coll*]	return true if *coll* contains no elements
includes[*coll*, *elem*]	return true if *elem* is a member of *coll*
fold[*coll*, v_0, *binop*]	iterate the binary function *binop* over *coll*, starting with v_0
select[*coll*, *crit*]	return a list of those elements that satisfy *crit*

Lists and sets provide a method for adding elements:

add[*list*, *elem*]	add *elem* to *list*

Indexed collections provide the following additional methods:

binaryDo[*icoll*, *binop*]	apply the binary function *binop* to all key-value pairs
includesKey[*icoll*, *key*]	return true if *key* is an existing key in *icoll*
at[*icoll*, *key*]	return the value with the given key
atPut[*icoll*, *key*, *val*]	insert a value with given key

Table 4.4–1: Methods for collections

The simplest collection to implement is `interval`. An interval is created with

$$\texttt{new[interval, } l, h, d\texttt{]}.$$

It consists of the integers $l, l + d, l + 2d, \ldots, h$ (in *Mathematica* this would be `Range[`l`, `h`, `s`]`).
Intervals are most often used to implement iteration in SMALLTALK. Note that we do not need to

```
Class[ collection, Object,
    {},
    {{fold,    Function[ {v0, binop}, Module[{v = v0},
                        do[self, (v = binop[v, #])&]; v] ]},
    {includes, Function[ {v},
         Catch[ do[self, If[#===v, Throw[True]]&]; False ] ]},
    {size,    fold[self, 0, #1+1&]&},
    {isEmpty, size[self]==0&},
    {select, Function[{crit},
             fold[self, new[list], (If[crit[#2], add[#1, #2]]; #1)&] ]}
    }
]
```

The abstract base class <code>collection</code>

generate and store all the values in an interval. We only need to implement the iteration method
do and the constructor. The iteration can simply be performed with the *Mathematica* function Do.
The constructor initializes the three instance variables.

```
Class[ interval, collection,
    {lower, upper, step},
    {{do,  Function[{f}, Do[ f[i], {i, lower, upper, step}]]},
     {new, (new[super]; {lower, upper, step}={##})&}
    }
]
```

The collection <code>interval</code>

Intervals are most often used to implement iteration in SMALLTALK. Here is a loop, SMALLTALK-style, that prints the squares of the first five integers.

```
In[1]:= do[ new[interval, 1, 5, 1], Print[#^2]& ]
1
4
9
16
25
```

The abstract base class for indexed collections implements some of the additional methods defined
for indexed collections. There is a new primitive method binaryDo[*coll*, *binop*] that needs to
be implemented in all subclasses. Other primitive methods are atPut[*coll*, *key*, *val*] and
atIfAbsent[*coll*, *key*, *proc*]. The latter returns the element stored under the given key; if
there is no such key it evaluates the parameterless procedure *proc* instead. Using these primitive
methods we can implement the other methods do, includesKey, and at.

The method do uses binaryDo and applies the unary function *f* to the second element of the
key-value pairs, ignoring the keys. includesKey tries to get the value stored at the given key. If
it succeeds, it returns True, otherwise the exception procedure is used to return immediately with
False. at uses atIfAbsent and prints an error message if the given key does not exist.

The simplest indexed collection is array. The keys are the integers from 1 up to the size of
the array. An array is created with new[array, *size*, v_0]. The new array is filled with copies

```
Class[ indexedCollection, collection,
    {},
    {{do, Function[{f}, binaryDo[self, f[#2]&]]},
     {includesKey, Function[{key},
                     Catch[atIfAbsent[self, key, Throw[False]&]; True] ]},
     {at, Function[{key},
            atIfAbsent[self, key, Message[indexedCollection::nsk, key, self]&] ]}
    }
]
```

The abstract base class <code>indexedCollection</code>

of v_0. In `array` we override the default implementation of `size` (from class `collection`) by a much faster method. We can compute the size of the array by simply looking at the size of the *Mathematica* list used to hold the elements. There is one private method `boundCheck` to test whether a given key is in range. `binaryDo` uses an interval to supply the key values. Much of this code could be written more efficiently. Our goal here is to emphasize code reuse rather than efficiency.

```
Class[ array, indexedCollection,
    {ar},
    {{binaryDo, Function[{f},
                  do[ new[interval, 1, size[self], 1],
                      f[#, at[self, #]]& ] ]},
     {atIfAbsent, Function[{key, exc},
                  If[ boundCheck[self, key], ar[[key]], exc[] ]
                  ]},
     {atPut, Function[{key, val},
                  If[ boundCheck[self, key],
                      ar[[key]] = val,
                      Message[array::nsk, key, size[self]] ];
                  self ]},
     {size,   Length[ar]&},
     {boundCheck, Function[ ind, TrueQ[1 <= #1 <= size[self]] ]},
     {new,    (new[super]; ar = Table[#2, {#1}])&}
    }
]
```

The indexed collection <code>array</code>

We create a new array of size 10, filled with 0.	`In[2]:= a = new[array, 10, 0]`
	
	`Out[2]= -array-`

We iterate over the array and store the value i^2 in element a_i.	`In[3]:= binaryDo[ a,` `              Function[{key, val}, atPut[a, key, key^2]] ]`

This returns a_8.	`In[4]:= at[a, 8]`
	
	`Out[4]= 64`

This computes the sum of all ele-   `In[5]:= fold[a, 0, Plus]`
ments.

`Out[5]= 385`

Lists and *sets* are implemented in terms of an auxiliary data type, the *link*. A link consists of
three parts: a key, a value, and a field to store the next link. This implements the familiar *linked
lists*. Links can be implemented as indexed collections. The code is not shown here. A list is a
collection with one instance variable, the list head, holding a link. To add an element, we create
a new link, and to iterate over a list we iterate over the links. Note that the key fields of the links
are ignored, since lists are not indexed collections.

```
Class[ list, collection,
    {head},
    {{do,     Function[{f}, binaryDo[head, f[#2]&]]},
     {add,    (head = new[link, 0, #1, head]; self)&},
     {new,    (new[super]; head=nullLink)&}
    }
]
```

The collection <code>list</code>

Sets can be implemented as lists. The only difference is that a certain element can only be included
once. Therefore we override the method `add` to check first whether the new element is already
present.

```
Class[ set, list,
    {},
    {{add, (If[!includes[self, #1], add[super, #1]]; self)&}
    }
]
```

The collection <code>set</code>

Lists are used by the method `select` to return the result of the selection.

Here we select all odd elements of   `In[6]:= select[ a, OddQ ]`
the array `a`.

`Out[6]= -list-`

The result is a list, which we look at   `In[7]:= do[ %, Print ]`
by iterating the print function.

```
81
49
25
9
1
```

The one remaining collection, `dictionary`, is the most complicated to implement. A dictionary
consists of key-value pairs, where keys and values can be arbitrary expressions. Storage and
retrieval of such arbitrary keys is best done with *hashing*. A hash function is a function that

computes from any expression an integer value in a certain range. Such a function is built into *Mathematica* as Hash[*expr*]. A dictionary can be implemented as an array of *bins*. Each bin is a linked list of key-value pairs. To insert a key-value pair, the hash value of the key is used to select one of the bins. We then traverse this linked list to check if the given key is already present. If it is we overwrite the value field, otherwise we add a new link. This implementation of dictionaries therefore makes heavy use of other collections (arrays and links) already defined. Immediate use of newly defined classes in the definition of other classes is typical of object-oriented programs. The class definition for dictionaries becomes amazingly short.

```
Class[ dictionary, indexedCollection,
    {linktable},
    {{binaryDo, Function[ {f}, do[linktable, binaryDo[#, f]&] ]},
     {atIfAbsent, Function[{key, exc},
                    atIfAbsent[ at[linktable, bin[self, key]],
                                 key, exc ] ]},
     {atPut, Function[{key, val}, With[{slot = bin[self, key]},
                atPut[ linktable, slot,
                        atPut[at[linktable, slot], key, val] ];
                self ]]},
     {new, (new[super]; linktable = new[array, #1, nullLink])&},
     {bin, Function[{key}, Mod[Hash[key], size[linktable]]+1]}
    }
]
```

The indexed collection <code>dictionary</code>

Observe how binaryDo is implemented. It simply iterates over the array of bins and calls binaryDo on the links in the table. The private method bin computes the bin to use for a given key. The size of the array of bins is determined at creation time and should be chosen depending on the expected size of the dictionary. A larger array speeds access but wastes some memory if there are only a few elements in the dictionary. A smaller array means slower access since the linked lists will be longer (remember that they are searched sequentially).

Here we create a new dictionary with 7 bins.

```
In[8]:= pb = new[dictionary, 7]

Out[8]= -dictionary-
```

We insert some phone numbers into it. The key is the name, the phone number is the value.

```
In[9]:= atPut[pb, "Wolfram Research", "(217) 398-0700"];\
        atPut[pb, "Miller Freeman", "(415) 905-2200"];\
        atPut[pb, "Academic Press", "(617) 876-3901"];\
        atPut[pb, "Apple Computer", "(408) 996-1010"];\
        atPut[pb, "DEC", "(508) 874-3111"];\
        atPut[pb, "IBM", "(800) 426-7378"];\
        atPut[pb, "NeXT", "(800) 848-NeXT"];\
        atPut[pb, "Sun Microsystems", "(800) USA-4SUN"];\
        atPut[pb, "Hewlett-Packard", "(800) 637-7740"];\
        atPut[pb, "Silicon Graphics", "(415) 960-1980"];\
        atPut[pb, "CONVEX", "(214) 497-4000"];
```

This looks up one of the numbers.

```
In[10]:= at[ pb, "Wolfram Research" ]

Out[10]= (217) 398-0700
```

Here we print the dictionary's contents. You may want to think about what the inner `fold` does.

```
In[11]:= binaryDo[ pb,
              Print[
                  #1,
                  fold[ new[ interval,
                          StringLength[#1], 20, 1 ],
                      "", #<>" "&],
                  #2 ]&
              ]
Apple Computer         (408) 996-1010
Hewlett-Packard        (800) 637-7740
Wolfram Research       (217) 398-0700
Academic Press         (617) 876-3901
NeXT                   (800) 848-NeXT
IBM                    (800) 426-7378
Miller Freeman         (415) 905-2200
Silicon Graphics       (415) 960-1980
CONVEX                 (214) 497-4000
DEC                    (508) 874-3111
Sun Microsystems       (800) USA-4SUN
```

With some "hacking" we can find the size of the bins.

```
In[12]:= do[ pb[[1]], Print[size[#]]& ]
2
3
1
3
0
1
1
```

4.5　Conclusions and Further Reading

Object-oriented programming is a useful tool for organizing larger programming projects or for modeling domains that are richly structured. We have shown how to implement in *Mathematica* a simple object-oriented system that provides for classes, (single) inheritance, and message passing.

The first object-oriented language was SIMULA, described in [3]. An excellent introduction to object-oriented programming is [6]. It discusses concepts and gives example programs in four different object-oriented languages: SMALLTALK, C++, Object Pascal, and Objective C. Other recommended books are [8], [36], and [4]. One of the most popular object-oriented languages is SMALLTALK, described in [16].

 The floppy disk contains the packages Classes.m, Accounts.m, AccountObjects.m, and Collections.m.

4.6   The Complete Code of Classes.m

```
BeginPackage["Classes`"]

Class::usage = "Class[class, superclass, variables, methods] defines a new class
    as a subclass of superclass. Class[object] gives the class of an object."
ClassQ::usage = "ClassQ[symbol] is True, if symbol is a class."
Methods::usage = "Methods[class] gives the list of methods of class."
InstanceVariables::usage = "InstanceVariables[class] gives the
    list of instance variables of class."
SuperClass::usage = "SuperClass[class] gives the superclass of class."
new::usage = "new[class, args...] generates a new object of class. Any
    method with name 'new' is called to initialize the new object."
delete::usage = "delete[obj] deletes an object (for special occasions only)."
self::usage = "self denotes the object inside methods."
super::usage = "super denotes the object as a member of its superclass."
isa::usage = "isa[obj, class] is true if obj belongs to class or a subclass of it."
NIM::usage = "NIM[self, <method>]& can be used as body of a pure virtual method."

Object::usage = "Object is the root class."

Object::nim = "Method `1` not implemented for class `2`."

Begin["`Private`"]

context = $Context

(* private rules for instances *) {variables}
(* private class methods *)       {methodHandler}
(* other private symbols *)       {raise}

ClassQ[_] := False (* default *)

Class[ class_Symbol,
       superclass_?ClassQ,
       variables:{_Symbol...},
       methods:{{_Symbol, _Function}...}|{}
     ] :=
    Module[{apply, standard,
            localmethods, allvariables, messages,
            methodQ},

        standard = { (* standard methods *)
            {Class,    class&},
            {isa, (class===#1 || isa[super, #1])&} };
        localmethods = Join[standard, methods];
        messages = Union[ Join[First /@ localmethods,
                               Methods[superclass]] ];
        allvariables = Join[variables, InstanceVariables[superclass]];
        (* class methods *)
        class/: Methods[class] = messages;
        class/: InstanceVariables[class] = allvariables;
        class/: SuperClass[class] = superclass;
```

```mathematica
      With[{ivars = Hold @@ allvariables,
            nvars = -Length[allvariables],
            localnames = First /@ localmethods,
            cookie = ToExpression[ context <> ToString[class] ],
            allvariables = allvariables},

        (* the head used for objects of this class *)
        SetAttributes[cookie, HoldAll];

        (* aux predicate for apply and message passing *)
        (methodQ[#] = True)& /@ messages;
        methodQ[_] = False;

        (* definitions of aux method applicator ``apply'' *)
        Apply[
          (apply[#1, obj_] :=
            With[{lvars = Take[Hold @@ obj, nvars]},
              #2 /.
                List @@ Thread[ ivars :> lvars, Hold] /.
                { self -> obj, super -> raise[obj, superclass] }
            ])&,
          localmethods, {1} ];
        (* inheritance, if not local method *)
        apply[f_, obj_] := methodHandler[superclass][f, obj];
        class/: methodHandler[class] = apply;

        (* message passing *)
        cookie/: (f_Symbol?methodQ)[obj_cookie, args___] :=
                apply[f, obj][args];
        (* super *)
        raise/: (f_Symbol?methodQ)[raise[obj_, class], args___] :=
                apply[f, obj][args];

        (* create instances of this class *)
        class/: new[class, init___] :=
          Module[{obj, syms = Unique[allvariables]},
        SetAttributes[Evaluate[syms], Temporary];
            obj = cookie @@ syms;
            new[obj, init]; (* call any constructor defined *)
            obj
          ];
        (* formatting *)
        Format[obj_cookie] := StringForm["-`1`-", class]
      ];
      class/: ClassQ[class] = True; (* seal of approval *)
      class
    ]

(* the class Object: we need a dummy superclass *)

Block[{noClass},
  noClass/: Methods[noClass] = {};
  noClass/: InstanceVariables[noClass] = {};
  noClass/: methodHandler[noClass] = badmessage;
```

```
  noClass/: SuperClass[noClass] = noClass;
  noClass/: ClassQ[noClass] = True;

  Class[ Object, noClass, {},
    {{new,                self&},
     {delete,             (Remove @@ self; Null)&},
     {isa,                #1===Object&},
     {Methods,            Methods[Class[self]]&},
     {InstanceVariables,  InstanceVariables[Class[self]]&},
     {SuperClass,         SuperClass[Class[self]]&},
     {NIM,          Message[Object::nim, #1, Class[self]]&}
    }
  ]
]

End[]

Protect[ Class, Methods, InstanceVariables, SuperClass,
        new, delete, isa, Object ]

EndPackage[]
```

Chapter Five
Databases

Databases are the most important commercial application of computers. Everybody seems to collect data of some sort. We will look at *relational databases,* explain their mathematical foundation, and discuss a few design issues that arise in developing a database for an application. We present a package with the most important database functions. Concepts are explained with an example, a database of classical music recordings.

About the illustration overleaf:
The fourth iteration of the tetrahedral sponge. It is obtained by cutting out an octahedron from a tetrahedron, leaving four smaller tetrahedra in the corners. Those four tetrahedra are then subdivided in the same way. The code is in the package Tetrix.m.

This picture was motivated by the *Sierpinski sponge* shown in *Programming in Mathematica.* A ray traced version of it has been rendered in Plate 5.

5.1 Database Design

Databases are an example of a model of the real world. Real-world objects are called *entities*. Entities of different sorts have certain *relationships* between them. An entity is described by a number of *attributes*, the properties of an object that we need to store in the database. Here, for example, is a simple database of recordings of classical music. The entity *record* has attributes *record title*, *label*, and *record number*. The entity *recording,* which describes a recording of a certain work of music, has the attributes *work title* and *composer name*. The relationship between records and recordings is that each recording appears on a certain record. The first approach to organizing this database is simply to list the attributes, one recording per line, as shown in Table 5.1–1.

| | | | Music | |
record title	*label*	*recordNo.*	*work title*	*composer name*
Das Konzert November 1989	Sony	45830	Conc. for Piano and Orch. No. 1	Beethoven, L. v.
Das Konzert November 1989	Sony	45830	Symphony No. 7	Beethoven, L. v.
Horovitz at Carnegie Hall	RCA	7992-2	Conc. for Piano and Orch. No. 5	Beethoven, L. v.
Horovitz at Carnegie Hall	RCA	7992-2	Conc. No. 1 for Piano and Orch.	Tchaikovsky, P. I.
Beethoven's Symphonies	London	430400-2	Symphony No. 6	Beethoven, L. v.
Beethoven's Symphonies	London	430400-2	Symphony No. 7	Beethoven, L. v.
Beethoven: The Piano Conc.	Angel	63360	Conc. for Piano and Orch. No. 1	Beethoven, L. v.
Beethoven: The Piano Conc.	Angel	63360	Conc. for Piano and Orch. No. 4	Beethoven, L. v.
Beethoven: The Piano Conc.	Angel	63360	Conc. for Piano and Orch. No. 5	Beethoven, L. v.

Table 5.1–1: A list of recordings of classical music

Many database programs for personal computers operate in this simple way. It has an important drawback: There is *redundancy* in the database. The record title and label, for example, are uniquely determined by the record number. Such redundancy wastes storage space. More importantly, it can lead to inconsistent updates since changes have to be made at several places instead of only once. Another redundancy is present in the attributes *work title* and *composer name*. Each such pair determines one work of classical music and may appear several times (in different recordings of that work). The redundancy problem can be solved by splitting the database into three lists, as show in Table 5.1–2.

The relationship between records and recordings is a *one-to-many* relationship, since each record may contain several recordings, while each recording belongs to exactly one record. The relationship is established by including the record number as a *key* in the list of recordings. A key is an attribute whose value is unique among all entries. The relationship between works and recordings is also one-to-many. Again we use a key attribute. Since there is no natural attribute that can serve as this key, we include a new attribute, the *workKey*. It is an arbitrary value (here we use a symbol) that is unique for each work.

Optional information can be incorporated into the design by splitting our lists still further. There can be any number of *artists* that perform a given work. Rather than have attributes $artist_1$,

Records		
record title	*label*	*record number*
Das Konzert November 1989	Sony Classical	45830
Horovitz at Carnegie Hall	RCA	7992-2
Beethoven's Symphonies	London	430400-2
Beethoven: The Piano Conc.	Angel	63360

Recordings		
recKey	*workKey*	*record number*
rec1	work1	45830
rec2	work2	45830
rec3	work3	7992-2
rec4	work4	7992-2
rec5	work5	430400-2
rec6	work2	430400-2
rec7	work1	63360
rec10	work9	63360
rec11	work3	63360

Works		
workKey	*work title*	*composer name*
work1	Conc. for Piano and Orch. No. 1	Beethoven, L. v.
work2	Symphony No. 7	Beethoven, L. v.
work3	Conc. for Piano and Orch. No. 5	Beethoven, L. v.
work4	Conc. No. 1 for Piano and Orch.	Tchaikovsky, Piotr I.
work5	Symphony No. 6	Beethoven, L. v.
work9	Conc. for Piano and Orch. No. 4	Beethoven, L. v.

Table 5.1–2: Removing redundancy

<table>
<tr><td colspan="3" align="center">Performers</td></tr>
<tr><td>recKey</td><td>artist name</td><td>instrument</td></tr>
<tr><td>rec1</td><td>Barenboim, Daniel</td><td>Berliner Philharmoniker</td></tr>
<tr><td>rec1</td><td>Barenboim, Daniel</td><td>Piano</td></tr>
<tr><td>rec2</td><td>Barenboim, Daniel</td><td>Berliner Philharmoniker</td></tr>
<tr><td>rec3</td><td>Horovitz, Vladimir</td><td>Piano</td></tr>
<tr><td>rec3</td><td>Reiner, Fritz</td><td>RCA Victor Symphony</td></tr>
<tr><td>rec4</td><td>Horovitz, Vladimir</td><td>Piano</td></tr>
<tr><td>rec4</td><td>Reiner, Fritz</td><td>RCA Victor Symphony</td></tr>
<tr><td>rec5</td><td>Solti, Sir Georg</td><td>Chicago Symphony Orch.</td></tr>
<tr><td>rec6</td><td>Solti, Sir Georg</td><td>Chicago Symphony Orch.</td></tr>
<tr><td>rec7</td><td>Barenboim, Daniel</td><td>Piano</td></tr>
<tr><td>rec7</td><td>Klemperer, Otto</td><td>New Philharmonia Orch.</td></tr>
<tr><td>rec10</td><td>Barenboim, Daniel</td><td>Piano</td></tr>
<tr><td>rec10</td><td>Klemperer, Otto</td><td>New Philharmonia Orch.</td></tr>
<tr><td>rec11</td><td>Barenboim, Daniel</td><td>Piano</td></tr>
<tr><td>rec11</td><td>Klemperer, Otto</td><td>New Philharmonia Orch.</td></tr>
</table>

Table 5.1–3: Performers

$artist_2, \ldots, artist_n$, for some arbitrary n (that will invariably turn out to be too small sometime in the future), we set up a new list of *performers*. It describes the relationship "artist x performs on recording y." For this we need a key in the list of recordings as well. We have already included *recKey* in the list recordings in Table 5.1–2 for this purpose. The resulting list of performers is shown in Table 5.1–3.

Again, we have some undesirable redundancy. The solution is to use a separate list of *artists* and include only an artist key in the list of performers. The result is in Table 5.1–4.

The list of performers consists entirely of key pairs. This is because there is a *many-to-many* relationship between recordings and artists. There is no natural entity that corresponds to this list.

While such a division of our data model into several lists would be cumbersome to use by hand, it should be no problem for a computer if it is properly programmed. A mathematically sound basis for this kind of databases is given by the *relational database model*.

5.2   Relational Databases

Each attribute has a certain set of possible values. Often the values are strings over some alphabet (the ISO-Latin 1 character set, for example). A *relation* is a subset of a Cartesian product of sets. Its elements are called *tuples*. As a set, a relation cannot contain multiple copies of identical tuples

Artists		
artistKey	*artist name*	*instrument*
art1	Barenboim, Daniel	Berliner Philharmoniker
art2	Barenboim, Daniel	Piano
art3	Horovitz, Vladimir	Piano
art4	Reiner, Fritz	RCA Victor Symphony
art5	Solti, Sir Georg	Chicago Symphony Orch.
art6	Klemperer, Otto	New Philharmonia Orch.

Performers	
recKey	*artistKey*
rec1	art1
rec1	art2
rec2	art1
. . .	. . .
rec11	art6

Table 5.1–4: Artists and performers

and there is no implied ordering of the tuples. A *database* consists of a number of relations. The tuples consist of one value for each attribute. In our example the lists are relations. The list of records (see Table 5.1–2) contains four tuples. All three attributes, *record title*, *label*, and *record number,* are strings.

Let us look at some design issues for deriving a collection of relations for a given problem. We will continue with the database for recordings of classical music that we've been using, augmented by a few more attributes. The basic entity is the *record* (nowadays more likely a CD, but we do not care about the physical storage medium at this point). The attributes we would like to keep track of are *title, label,* and *record number.* A record typically contains a number of *recordings* of *works.* The work has a *title, opus number*, and a *composer.* The recording is made on a certain *date* by a number of *artists*, each of them playing an *instrument* (which may be their own voice). Often the orchestra is treated as the "instrument" of the conductor. A composer is described by name.

These simple relationships can be visualized in a diagram of hierarchical data-dependency, another frequently used description of a database (Figure 5.2–1).

The relationships described are indicated by *arcs* between two entities. The arc from records to recordings is labeled 1—n, indicating that one record may contain many recordings, but each recording belongs to exactly one record (a *one-to-many* relationship). The relationship between recordings and artists is *many-to-many*, indicated by n—m. Since we have already removed redundancy by splitting our data in a hierarchical fashion, the design of the necessary relations becomes rather easy. Each entity gives rise to one relation with the attributes mentioned earlier. A many-to-one relationship can be implemented with a key, as we have seen. Therefore we need keys for records, works, and composers. A many-to-many relationship would again introduce

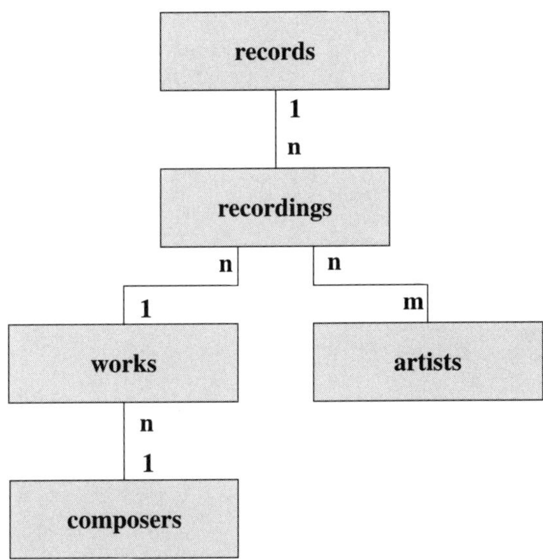

Figure 5.2–1: Basic entities and relationships in the music database

duplicate information, so we define an additional relation that holds only the keys of the related entities. We introduce the additional relation *performers* between recordings and artists.

The resulting design is shown in Table 5.2–1. The starred attributes are keys for their relation. The *type* column lists the data type used to store values of the corresponding attributes. They will be used for data entry. The type *expression* means any *Mathematica* expression (we will use symbols only for the most part).

Exercise: Put this information itself into a database. What are its relations and attributes? This is called the *meta-database*.

We can implement a relation as an abstract data type having the following representation:

$$\texttt{relation[\{}\textit{attributes}\ldots\texttt{\}, \{}\textit{tuples}\ldots\texttt{\}]}\,.$$

The first element is the list of attributes, the second element is the list of tuples. Each tuple is again represented as a list of values. The two components can be extracted with the selectors `Fields[`*relation*`]` and `Tuples[`*relation*`]`. For obvious reasons we cannot name the selector `Attributes`. The term *field* is often used in place of attribute. The constructor `newRelation[]` creates new relations. It uses `Union[]` on the tuples to enforce the restriction that no two tuples can be identical.

The package Databases.m reproduced at the end of this chapter implements the most important database functions. We will now look at some of these functions in more detail.

Music		
relation	*attribute*	*type*
records	recordNumber*	string
	label	string
	recordTitle	string
recordings	recKey*	expression
	workKey	expression
	date	string
	recordNumber	string
works	workKey*	expression
	compKey	expression
	workTitle	string
	opus	string
composers	compKey*	expression
	composerName	string
artists	artistKey*	expression
	artistName	string
	instrument	string
performers	recKey	expression
	artistKey	expression

Table 5.2–1: Relations and fields of the music database

5.3 Fundamental Operations on Relations

There are two fundamental types of operations on databases: *updates* and *queries*. An update modifies the information stored in the database, a query is used to select part of the information and to look at it. These operations translate into elementary operations on the relations that comprise the database. Let us look at queries first.

A *selection* returns a subset of the tuples of a relation, determined by a Boolean test applied to the tuples. We can conveniently overload `Select` to perform this function.

The code sets up the pure function needed for the selection itself. We can therefore express our selection function as a Boolean expression using the names of the attributes directly. The selection is then performed on the tuples using the built-in version of `Select`. For example, to select all records published by Sony Classical we could use

```
Select[records, label == "Sony Classical"].
```

```
Select[rel_relation, cond_] :=
   With[ {fields = Fields[rel], tuples = Tuples[rel]},
   Module[{pred, selt},
      pred = Function @@ {fields, cond};
      selt = Select[ tuples, Apply[pred, #]& ];
      newRelation[ fields, selt ]
   ]]
```

Selecting tuples of a relation

Another important operation is *projection*. Projection deletes some of the attributes and returns a relation consisting of the remaining attributes only. We will use it frequently to get rid of unwanted attributes (keys, for example) before displaying relations. The syntax is

$$\texttt{Projection[}\textit{relation},\ \{\textit{field}_1,\ \textit{field}_2,\ \ldots\}\texttt{]}\ .$$

If a relation R has attributes A_1, A_2, $\ldots$, A_n, a projection onto attributes A_{i_1}, A_{i_2}, $\ldots$, A_{i_k} simply retains the elements numbered i_1, i_2, $\ldots$, i_k of all tuples. Such a projection is denoted by $\pi_{A_{i_1},A_{i_2},\ldots,A_{i_k}}(R)$. A simple implementation is to apply the pure function

$$\texttt{Function[}\{A_1,\ A_2,\ \ldots,\ A_n\},\ \{A_{i_1},\ A_{i_2},\ \ldots,\ A_{i_n}\}\texttt{]}$$

to all tuples.

```
Projection[rel_relation, subfields_List] :=
   With[ {fields = Fields[rel], tuples = Tuples[rel]},
   Module[{pred, newt},
      pred = Function @@ {fields, subfields};
      newt = Apply[pred, tuples, {1}];
      newRelation[ subfields, newt ]
   ]]
```

Projection

For different sorts of queries, for example "find all records that do not contain any works by Beethoven," we need a few more operations on relations. The *union* $r_1 \cup r_2$, *intersection* $r_1 \cap r_2$, and *complement* or difference $r_1 - r_2$ of relations is defined for relations with the same attributes. The operation is simply performed on their tuples.

5.3.1 The Natural Join

The most important primitive operation on relations is the *natural join*, which assembles two relations into a larger one. Assume that the relations r_1 and r_2 have one attribute in common. Let r_1 have the attributes A_1, A_2, $\ldots$, A_{n-1}, B, and r_2 have the attributes B, C_2, C_3, $\ldots$, C_n. The natural join $r_1 \bowtie r_2$ is a relation with attributes A_1, A_2, $\ldots$, A_{n-1}, B, C_2, C_3, $\ldots$, C_n. It consists of all tuples of the form $(a_1, a_2, \ldots, a_{n-1}, b, c_2, c_3, \ldots, c_n)$ such that $(a_1, a_2, \ldots, a_{n-1}, b) \in r_1$ and $(b, c_2, c_3, \ldots, c_n) \in r_2$. If r_1 and r_2 have more attributes in common, they must all agree. If they do not have an attribute in common, the natural join degenerates into the *Cartesian product*.

The implementation of the natural join gets rather messy. The main problem is to avoid large intermediate data. The size of the join could possibly be equal to the *product* of the sizes of the two relations. Normally it is only a small fraction of this maximum. For performance reasons we sort the two relations on the common attributes. In this way we can avoid repeated searching for matching attributes. The implementation in the package Databases.m has a runtime proportional to the actual size of the join.

If the attribute in common is a key that we introduced to split a list into two relations, the join undoes this split and returns the original list, with all redundancy. Often this is used for preparing printouts of databases. The join of the records and recordings gives a relation that lists all records and recordings.

<table>
<tr><td>

The package RecordManager.m defines a few auxiliary functions for working with the music database.

</td><td>

```
In[1]:= <<RecordManager.m
```

</td></tr>
<tr><td>

load reads in relations from the sample database music.db.

</td><td>

```
In[2]:= load
```

</td></tr>
<tr><td>

The join of the records and recordings gives a relation that lists all records and recordings. Note that we defined a print form for relations that lists the attributes explicitly, but not the tuples.

</td><td>

```
In[3]:= Join[records, recordings]
Out[3]= R[{recordTitle, label, recordNumber, recKey,
    workKey, date}, {<<22>>}]
```

</td></tr>
<tr><td>

The works are still identified only by a key. A further join with the works relation and the composers relation retrieves information about the works in a readable form.

</td><td>

```
In[4]:= Join[%, works, composers]
Out[4]= R[{recordTitle, label, recordNumber, recKey,
    workKey, date, compKey, workTitle, opus, composerName},
    {<<22>>}]
```

</td></tr>
<tr><td>

The key fields are no longer needed and can be projected away, along with other unwanted information.

</td><td>

```
In[5]:= Projection[ %, {recordTitle, recordNumber,
                         composerName, workTitle, opus} ]
Out[5]= R[{recordTitle, recordNumber, composerName,
    workTitle, opus}, {<<22>>}]
```

</td></tr>
<tr><td>

Here is a quick look at the result.

</td><td>

```
In[6]:= Short[ Tuples[%], 4 ]
Out[6]//Short=
  {{Beethoven's Symphonies, 430400-2, Beethoven, L. v.,
    Symphony No. 1, 21}, <<20>>,
    {Schubert, Symphonies Nos. 1&2, Wand, CDC 7 47874 2,
    Schubert, Franz, Symphony No. 2, 125}}
```

</td></tr>
</table>

This form of output is not good enough. Instead, we want a tabular form called a *report*. Indentation is used in place of repeating attributes whose values are the same from one line to the next.

The numbers n_i given in the command Report[*relation*, n_1, ...] specify the number of attributes that belong together when comparing one tuple with the next one.

```
In[7]:= Report[ %%, 2, 1, 2 ];
Beethoven's Symphonies, 430400-2
    Beethoven, L. v.
        Symphony No. 1, 21
        Symphony No. 2, 36
        Symphony No. 3, 55
        Symphony No. 4, 60
        Symphony No. 5, 67
        Symphony No. 6, 68
        Symphony No. 7, 92
        Symphony No. 8, 93
        Symphony No. 9, 125
Beethoven: The Piano Concertos, 63360
    Beethoven, L. v.
        Conc. for Piano and Orch. No. 1, 15
        Conc. for Piano and Orch. No. 2, 19
        Conc. for Piano and Orch. No. 3, 37
        Conc. for Piano and Orch. No. 4, 58
        Conc. for Piano and Orch. No. 5, 73
Das Konzert November 1989, 45830
    Beethoven, L. v.
        Conc. for Piano and Orch. No. 1, 15
        Symphony No. 7, 92
Horovitz at Carnegie Hall, 7992-2
    Beethoven, L. v.
        Conc. for Piano and Orch. No. 5, 73
    Tchaikovsky, Piotr I.
        Conc. No. 1 for Piano and Orch., 23
Mussorgsky/Tchaikovsky, GD 60 449 QH
    Mussorgsky, M.
        Pictures at an Exhibition,
    Tchaikovsky, Piotr I.
        Conc. No. 1 for Piano and Orch., 23
Schubert, Symphonies Nos. 1&2, Wand, CDC 7 47874 2
    Schubert, Franz
        Symphony No. 1, 82
        Symphony No. 2, 125
```

Here's another example. We would like a full listing of the contents of the CD "Das Konzert November 1989," performed in Berlin after the opening of the Berlin Wall.

First we select from the recordings the ones that are on this disk (whose number is 45830).

```
In[8]:= Select[ recordings, recordNumber == "45830" ]

Out[8]= R[{recKey, workKey, date, recordNumber},

    {{recKey$16, workKey$14, 1989.11.12, 45830}, <<1>>}]
```

Now we join it with all needed relations and project onto the desired attributes.

```
In[9]:= Join[ %, works, composers, performers, artists ]

Out[9]= R[{recKey, workKey, date, recordNumber, compKey,

    workTitle, opus, composerName, artistKey, artistName,

    instrument}, {<<3>>}]

In[10]:= Projection[ %, {composerName, workTitle,
                         artistName, instrument} ];
```

Finally, we generate a report.

```
In[11]:= Report[ %, 1, 1, 2 ];

Beethoven, L. v.
    Conc. for Piano and Orch. No. 1
        Barenboim, Daniel, Berliner Philharmoniker
        Barenboim, Daniel, Piano
    Symphony No. 7
        Barenboim, Daniel, Berliner Philharmoniker
```

Let us now find the records that do not contain any Beethoven.

First, here is a relation that contains the composers and record numbers of all records.

```
In[12]:= rc =
            Projection[
                Join[records, recordings, works, composers],
                {recordNumber, composerName} ]
Out[12]= R[{recordNumber, composerName},

    {{CDC 7 47874 2, Schubert, Franz}, <<7>>}]
```

To retrieve the records that do contain works by Beethoven we use an auxiliary relation containing just one tuple.

```
In[13]:= b = newRelation[ {composerName},
                        {{"Beethoven, L. v."}}];
```

Since we need only their record numbers, we use a projection after the natural join.

```
In[14]:= br = Projection[ Join[ rc, b ], {recordNumber} ]
Out[14]= R[{recordNumber},

    {{430400-2}, {45830}, {63360}, {7992-2}}]
```

The projection of the records relation onto the record number lists *all* records.

```
In[15]:= ar = Projection[ records, {recordNumber} ]
Out[15]= R[{recordNumber},

    {{CDC 7 47874 2}, {GD 60 449 QH}, <<3>>, {7992-2}}]
```

We compute the complement to get those records that do not contain any Beethoven.

```
In[16]:= Complement[ ar, br ]
Out[16]= R[{recordNumber},

    {{CDC 7 47874 2}, {GD 60 449 QH}}]
```

A join obtains more information about these records to report.

```
In[17]:= Join[ %, records, recordings, works, composers ];
```

The projection gets rid of all key fields.

```
In[18]:= Projection[ %, {recordTitle, label,
                        recordNumber, composerName} ]
Out[18]= R[{recordTitle, label, recordNumber,

    composerName}, {<<3>>}]
```

Here, finally, is the result.

```
In[19]:= Report[ %, 3, 1 ];

Mussorgsky/Tchaikovsky, RCA, GD 60 449 QH
    Mussorgsky, M.
    Tchaikovsky, Piotr I.
Schubert, Symphonies Nos. 1&2, Wand, EMI, CDC 7 47874 2
    Schubert, Franz
```

5.4 Data Entry

We have seen how to select and display information contained in a database. How does it get there in the first place? It is clearly not feasible to modify the tuples of the relations directly. The data entry needs to be modeled after the logical organization of the database, as described earlier and displayed in Figure 5.2–1. The program should then create the necessary key attributes and modify the relations in the required way. We will discuss a program to enter new information into our music database after a brief look at the underlying primitive operations.

The lowest level operation is `addTupleTo[`*relation*`, `*tuple*`]`, which adds one tuple to a relation. Like the built-in operations `AppendTo`, it is a destructive operation that overwrites the value of the variable holding the relation. Therefore it needs the attribute `HoldFirst`. The tuple is added with the help of `Union` to avoid duplicate entries.

```
SetAttributes[addTupleTo, HoldFirst]
addTupleTo[ rel_Symbol, tuple_List ] :=
   SetTuples[rel, Union[Tuples[rel], {tuple}]] /; RelationQ[rel]

SetAttributes[SetTuples, HoldFirst]
SetTuples[rel_Symbol, newtuples_] :=
   (rel = ReplacePart[ rel, newtuples, 2 ]) /; RelationQ[rel]
```

Adding a tuple to a relation

If one of the attributes is a key, we want to let the program generate a new unique value for it rather than entering a value ourselves. This value is then inserted into the tuple in the correct position and the complete tuple is added to the relation.

```
SetAttributes[addWithKeyTo, HoldFirst]

addWithKeyTo[ rel_Symbol, tuple_List, key_Symbol ] :=
   With[ {fields = Fields[rel], keyval = Unique[key]},
   Module[{kpos, nt},
      kpos = Position[ fields, key ][[1,1]];
      nt = Insert[ tuple, keyval, kpos ];
      addTupleTo[ rel, nt ];
      keyval
   ]] /; RelationQ[rel] && MemberQ[Fields[rel], key]
```

Adding a tuple to a relation with a key attribute

Finally, we need routines for the data entry itself. The basic interactive input operation is `InputField[`*attr*`, `*type*`, `*prompt*`]`, which prompts for input of one value of the given type. A good implementation has safeguards against bad values and also provides an escape mechanism. We chose to use an end-of-file mark to abort data entry. On many computers it is generated by CNTRL-D (the familiar "hold down the control key while hitting D"). Under the notebook front end or on machines where CNTRL-D doesn't work, you can enter a dot to abort input. Under the front end, input is requested through panels. This doesn't work quite as well as under the kernel alone.

When entering data for new records it will often happen that a work, an artist, or a composer is already in the database. Therefore we should first ask for those attributes that determine an entry completely and then search the database for that entry. If it is found we do not ask for the remaining attributes. For example, we first ask for the record number and then search the relation of records. Only if this record is not already there do we ask for its title and label. For details on the operation of this higher level input function, called `FindOrAdd[ ]`, please consult the package Databases.m.

The data entry procedure is a series of nested loops. The outer one asks for the next record to enter. Then we loop over the recordings on this record. We determine the work performed by asking for its title and the composer. Then we loop over the performers by repeatedly asking for the next artist and instrument. Escaping from the loop is done with CNTRL-D as described previously. Here is a sample session to enter the data from a CD containing two of Mozart's violin concertos.

```
In[19]:=  DataEntry
```
recordNumber: 410 020-2
label: Deutsche Grammophon
recordTitle: Mozart Violin Concertos Nos. 3&5
next work on record 410 020-2
   composerName: Mozart, W. A.
   workTitle: Conc. for Violin and Orch. No. 3
   opus: 216
   Recording date: 1983
   next performer of Conc. for Violin and Orch. No. 3
      artistName: Perlman, Itzhak
      instrument: Violin
   next performer of Conc. for Violin and Orch. No. 3
      artistName: Levine, James
      instrument: Wiener Philharmoniker
   next performer of Conc. for Violin and Orch. No. 3
      artistName: ∧D
next work on record 410 020-2
   composerName: Mozart, W. A.
   Found: `{compKey$16, Mozart, W. A.}`
   workTitle: Conc. for Violin and Orch. No. 5
   opus: 219
   Recording date: 1983
   next performer of Conc. for Violin and Orch. No. 5
      artistName: Perlman, Itzhak

    instrument: Violin
    Found: {artistKey$31, Perlman, Itzhak, Violin}
  next performer of Conc. for Violin and Orch. No. 5
    artistName: Levine, James
    instrument: Wiener Philharmoniker
    Found: {artistKey$37, Levine, James, Wiener Philharmoniker}
  next performer of Conc. for Violin and Orch. No. 5
    artistName: ^D
next work on record 410 020-2
  composerName: ^D

recordNumber: ^D

In[20]:= **save**

One major source of problems still remains: misspelling of names. A production-quality data entry system should deal with such errors. *Mathematica* provides a string comparison function that allows for spelling errors: StringMatchQ[*string*, *pattern*, SpellingCorrection->True]. It could be used to implement a more sophisticated data entry program. This only scratches the surface, however. Names of Russian composers (like *Tchaikovsky* in our example) are transliterated from the Cyrillic original in many different ways. This is a constant source of frustration.

    The package RecordManager.m contains all the code needed to maintain the records database. The function init creates an empty database, load loads data from a file music.db and save writes back the possibly updated information. A number of standard reports have also been defined.

5.5 Other Functions

Our database manager Databases.m contains a few more operations. We have overloaded the iterators Do, Table, and Sum to allow for a relation as an iterator. The expression iterated over may contain attribute names. They serve as iterator variables and successively get their values from all tuples in the relation. For example, this is the code for Do:

```
Do[ body_, rel_ ] :=
   With[ {fields = Fields[rel], tuples = Tuples[rel]},
      Module[fields,
         Do[ fields = tuples[[i]]; body, {i, Length[tuples]} ]
      ]
   ] /; RelationQ[rel]
```

Iterating over a relation

This type of iteration gives us another way to display relations. Here are the titles of all records in our database.

```
In[21]:= Do[ Print[recordTitle], records ]

Beethoven: The Piano Concertos
Beethoven's Symphonies
Das Konzert November 1989
Horovitz at Carnegie Hall
Mussorgsky/Tchaikovsky
Schubert, Symphonies Nos. 1&2, Wand
```

In a simple inventory database we could use Sum to find the total value of all goods in a store. Suppose the relation `items` contains the attributes *itemNo*, *itemName*, *cost*, and *price*, while the `inventory` relation contains attributes *itemNo* and *count*. The value of the items in the inventory could then be computed simply as

$$\texttt{Sum[count*cost, Join[items, inventory]]}.$$

Another iterator `Modify[`*body*, *relation*`]` writes back modified values after each iteration. With it we could, for example, increase the price of our items by 10 percent in this simple way:

$$\texttt{Modify[price = 1.1 price, items]}.$$

5.6 Conclusions

We have looked at only a few issues relating to databases. Let us briefly mention some other topics.

- *Physical data storage:* Larger databases do not fit into the main memory of a computer as do our sample relations. Data are stored on disks and only a small fraction resides in fast memory. Efficient organization of disk access is therefore of primary concern for implementing queries and updates. One tool for this is an *index*. An index into a relation is a small table that lists for each occurring key-value the location of the tuple with this key-value. Since key fields are often used in a natural join, an index speeds up this operation considerably.

- *Consistency:* A *transaction* is a sequence of related primitive update operations. In our example, entering the data for a whole record would be a transaction. If the sequence of primitive updates is interrupted in the middle (because of a hardware or software failure or because of a CNTRL-D in the wrong place in our `DataEntry` procedure) the database is left in an *inconsistent state*. Database systems therefore provide for means to ensure that a transaction either *completes* successfully or can be completely *undone*.

- *Concurrency:* During a transaction the database is temporarily in an inconsistent state. Real-world databases may be accessed by thousands of users (humans or programs) at the same time. Those parts that are inconsistent must be *locked* from access by others. Access is allowed again only after the transaction has finished.

The operations on relations that we presented (join, union, intersection, etc.) are part of the *relational algebra*.dxAlgebra!relational Another set of primitive operations, called the *relational calculus*, is often used. Its primitive expressions are predicates formed with variables that range over tuples. It is the basis for SQL, an industry standard query language. In SQL, our query to list the contents of one particular CD (see lines In[8] through In[11]) would look like this:

```
select composerName, workTitle, artistName, instrument
    of recordings, works, composers, performers, artists
    where recordings.recordNumber = "45830" and
          recordings.workKey = works.workKey and
          works.compKey = composers.compKey and
          recordings.recKey = performers.recKey and
          performers.artistKey = artists.artistKey
go
```

As you can see, projection, selection, and joins can be combined. This is quite natural to use but not as easy to implement as our relational algebra. In commercial database programs you will often find the terms *table* instead of relation and *row* instead of tuple. More on all aspects of databases can be found in Ullman's excellent book [45]. The musical data were taken from the American *Schwann Opus* catalog and the German *Bielefelder Katalog Klassik*.

 The floppy disk contains the packages **Databases.m** and **RecordManager.m**, as well as the sample database **music.db**.

5.7   The Complete Code of Databases.m

```
BeginPackage["Databases`"]

(* data type *)

newRelation::usage = "newRelation[fields, tuples] creates a new relation."
Fields::usage = "Fields[relation] gives the list of fields (attributes)."
Tuples::usage = "Tuples[relation] gives the tuples of a relation."
Type::usage = "Type[attr] gives the type of the attribute. The default is String."
Prompt::usage = "Prompt[attr] gives the prompt to use for input. The default is 'attr: '."

(* basic operations of relational calculus *)

Projection::usage = "Projection[rel, {fields..}] projects the relation to
    the given fields which must be a subset of its fields."
Join::usage = Join::usage <>
    " Join[rel1, rel2] gives the natural join of the relations rel1 and rel2."
```

```
Complement::usage = Complement::usage <>
    " Complement[rel1, rel2] gives the difference of the relations rel1 and rel2."
Quotient::usage = Quotient::usage <>
    " Quotient[rel1, rel2] gives the quotient of the relations rel1 and rel2."
Union::usage = Union::usage <>
    " Union[rel1, rel2] gives the union of the relations rel1 and rel2."
Intersection::usage = Intersection::usage <>
    " Intersection[rel1, rel2] gives the intersection of the
    relations rel1 and rel2."

(* selection *)

Select::usage = Select::usage <>
    " Select[ rel, cond ] selects all tuples in rel that satisfy cond."
FindOne::usage = "FindOne[rel, cond] returns one tuple in rel that satisfies
    the given condition, or $Failed if none exists."

(* update functions *)

addTupleTo::usage = "addTupleTo[rel, tuple] adds tuple to the relation rel."
addWithKeyTo::usage = "addWithKeyTo[rel, tuple, key] adds the tuple together
    with a unique key to rel. It returns the key generated."
Modify::usage = "Modify[body, rel] loops over the tuples of rel and
    assigns any changed values back to rel."

(* interactive input *)

InputField::usage = "InputField[attr, type, prompt] asks for input.
    The defaults for type and prompt are Type[attr] and Prompt[attr]."
FindOrAdd::usage = "FindOrAdd[rel, attrs, presets, key] is a higher-level
    input function."

(* output *)

Report::usage = "Report[rel, l1, l2, ...] generates a report with li fields
    on one line."

(* misc *)

RelationQ::usage = "RelationQ[rel] gives True, if rel is a relation."
R::usage = "R[fields, Short[tuples]] is the printform for relations."


Begin["`Private`"]

subSetQ[set_List, subset_List] := Complement[subset, set] === {}

newRelation[fields:{___Symbol}, t_List:{}] := relation[ fields, Union[t] ] /;
    t === {} || TensorRank[t] >= 2 && Dimensions[t][[2]] == Length[fields]

Fields[relation[f_, t_, ___]] := f
Tuples[relation[f_, t_, ___]] := t

SetAttributes[SetTuples, HoldFirst]

SetTuples[rel_Symbol, newtuples_] :=
    (rel = ReplacePart[ rel, newtuples, 2 ]) /; RelationQ[rel]

RelationQ[rel_] := Head[rel] == relation && Length[rel] >= 2
```

```
protected = Unprotect[Do, Table, Sum];

relation /: Union[rel1_relation, rel2_relation] :=
    newRelation[ Fields[rel1], Union[Tuples[rel1], Tuples[rel2]] ] /;
        Fields[rel1] === Fields[rel2]

relation /: Intersection[rel1_relation, rel2_relation] :=
    newRelation[ Fields[rel1], Intersection[Tuples[rel1], Tuples[rel2]] ] /;
        Fields[rel1] === Fields[rel2]

relation /: Complement[rel1_relation, rel2_relation] :=
    newRelation[ Fields[rel1], Complement[Tuples[rel1], Tuples[rel2]] ] /;
        Fields[rel1] === Fields[rel2]

SetAttributes[addTupleTo, HoldFirst]

addTupleTo[ rel_Symbol, tuple_List ] :=
    SetTuples[rel, Union[Tuples[rel], {tuple}]] /; RelationQ[rel]

SetAttributes[addWithKeyTo, HoldFirst]

addWithKeyTo[ rel_Symbol, tuple_List, key_Symbol ] :=
    With[ {fields = Fields[rel], keyval = Unique[key]},
    Module[{kpos, nt},
        kpos = Position[ fields, key ][[1,1]];
        nt = Insert[ tuple, keyval, kpos ];
        addTupleTo[ rel, nt ];
        keyval
    ]] /; RelationQ[rel] && MemberQ[Fields[rel], key]

relation /: Select[rel_relation, cond_] :=
    With[ {fields = Fields[rel], tuples = Tuples[rel]},
    Module[{pred, selt},
        pred = Function @@ {fields, cond};
        selt = Select[ tuples, Apply[pred, #]& ];
        newRelation[ fields, selt ]
    ]]

FindOne[rel_relation, cond_] :=
    With[ {fields = Fields[rel], tuples = Tuples[rel]},
    Module[{pred, selt},
        pred = Function @@ {fields, cond};
        selt = Select[ tuples, Apply[pred, #]&, 1 ];
        If[ Length[selt] == 0, $Failed, selt[[1]] ]
    ]]

relation /: Projection[rel_relation, subf_List] :=
    With[ {fields = Fields[rel], tuples = Tuples[rel]},
    Module[{pred, newt},
        pred = Function @@ {fields, subf};
        newt = Apply[pred, tuples, {1}];
        newRelation[ subf, newt ]
    ]] /; subSetQ[Fields[rel], subf]
```

```
relation /: Join[rel1_relation, rel2_relation] :=
    With[ {f1 = Fields[rel1], f2 = Fields[rel2],
           t1 = Tuples[rel1], t2 = Tuples[rel2]},
      Module[{res = cons[], val1 = Null, p1, p2, lc, inv,
              pos1, pos2, ord1, ord2, record, high = 1, take, mult},
         common = Intersection[f1, f2];
         lc = Length[common];
         pos1 = Flatten[ Position[f1, #]& /@ common ];
         pos2 = Flatten[ Position[f2, #]& /@ common ];
         rest1 = Complement[ Range[Length[f1]], pos1 ];
         rest2 = Complement[ Range[Length[f2]], pos2 ];
         order1 = Join[pos1, rest1];
         order2 = Join[pos2, rest2];
         ord1 = Sort[ #[[order1]]& /@ t1 ];
         ord2 = Sort[ #[[order2]]& /@ t2 ];
         Do[
             If[ val1 =!= Take[ ord1[[p1]], lc ],
                 val1 = Take[ ord1[[p1]], lc ];
                 (* search occurrences in pos2 *)
                 low = high;
                 While[ low <= Length[ord2] &&
                        Order[val1, Take[ ord2[[low]], lc ]] == -1,
                        low++
                 ];
                 high = low;
                 While[ high <= Length[ord2] &&
                        Order[val1, Take[ ord2[[high]], lc ]] == 0,
                        high++
                 ];
                 (* low ... high-1 have same values *)
                 If[ high > low,
                     take = Take[ ord2, {low, high-1} ];
                     take = Drop[Transpose[take], lc]
                 ];
             ];
             (* form product *)
             If[ high > low,
                 mult = Table[#, {high-low}]& /@ ord1[[p1]];
                 record = Transpose[ Join[mult, take] ];
                 res = cons[res, ##]& @@ record
             ],
           {p1, 1, Length[t1]}];
         res = List @@ Flatten[res];
         If[ Length[res] > 0,
           inv = Flatten[ Position[order1, #]& /@ Range[Length[f1]] ];
           inv = Join[ inv, Length[f1] + Range[Length[f2]-lc] ];
           res = Transpose[ Transpose[res][[inv]] ];
         ];
         newRelation[ Join[ f1, f2[[rest2]] ], res ]
      ]]
```

```
relation /: Quotient[rel1_relation, rel2_relation] :=
    With[ {f1 = Fields[rel1], f2 = Fields[rel2]},
    Module[{fields, t, ts, v},
        fields = Complement[f1, f2];
        t =  Projection[rel1, fields];
        ts = Projection[ Join[t, rel2], f1 ];
        v =  Projection[ Complement[ts, rel1],  fields ];
        Complement[t, v]
    ]] /; subSetQ[Fields[rel1], Fields[rel2]]

Do[ body_, rel_ ] :=
    With[ {fields = Fields[rel], tuples = Tuples[rel]},
        Module[fields,
            Do[ fields = tuples[[i]]; body, {i, Length[tuples]} ]
        ]
    ] /; RelationQ[rel]

Table[ body_, rel_ ] :=
    With[ {fields = Fields[rel], tuples = Tuples[rel]},
        Module[fields,
            Table[ fields = tuples[[i]]; body, {i, Length[tuples]} ]
        ]
    ] /; RelationQ[rel]

Sum[ body_, rel_ ] :=
    Plus @@ Table[ body, rel ] /; RelationQ[rel]

SetAttributes[ Modify, HoldAll ]
Modify[ body_, rel_Symbol ] :=
    With[ {fields = Fields[rel]},
    Module[ {tuples = Tuples[rel], i},
        Module[fields,
            Do[ fields = tuples[[i]];
                body;
                tuples[[i]] = fields,
              {1, Length[tuples]} ]
        ];
        rel = newRelation[ fields, tuples ]
    ]] /; RelationQ[rel]

(* report generation *)

indent = 4
separ = ", "
Report[rel_relation, levels___] :=
    With[ {fields = Fields[rel], tuples = Tuples[rel]},
    Module[{i, lev = {levels}, next, old, i1},
        If[ Plus @@ lev < Length[fields],
            AppendTo[lev, Length[fields] - Plus @@ lev] ];
        nest = Length[lev];
        old = Table[ Null, {nest+1} ];
        Do[ i1 = 0;
            Do[
                range = Range[ i1 + 1, i1 + lev[[i]] ];
```

```
                    i1 += lev[[i]];
                    line = fields[[range]];
                    If[ line =!= old[[i]],
                        Print[ Spaces[indent(i-1)], Infix[line, separ] ];
                        old[[i]] = line;
                        old[[i+1]] = Null;
                    ];
                  , {i, nest}
                ];
              , rel
            ];
            rel
    ]]

Spaces[i_] := Nest[ # <> " "&, "", i ]

(* fix Infix problem for functions with one argument *)

iprot = Unprotect[Infix]
Infix[_[e_], h_:Null] := e
Protect[Evaluate[iprot]]

(* prompted input *)
(* ^D aborts (returns EndOfFile) *)
(* Default type is String *)

Type[_] := String

(* default prompt is fieldname *)

Prompt[field_] := ToString[StringForm["`1`: ", field]]

noinput = "end of input"
abort = "."

InputField[field_] := InputField[field, Type[field], Prompt[field]]
InputField[field_, String, prompt_] :=
    Module[{ans},
    ans = InputString[prompt];
    If[ ans === EndOfFile || ans === abort,
        Print[noinput]; Return[EndOfFile] ];
    ans
    ]

InputField[field_, type_, prompt_] :=
    Module[{str, strstream, res},
        While[ True,
        str = InputString[prompt];
        If[ str === EndOfFile || str === abort,
            Print[noinput]; Return[EndOfFile] ]; (* no input *)
        strstream = StringToStream[str];
        res = Read[ strstream, type ];
        Close[strstream];
        If[ res === EndOfFile || res === $Failed, (* error *)
            Print["Reenter Data"],
            Break[]
```

Plate 1: Divisor Plot
The number of divisors of the integers from 1 to 199^2 is shown in a spiral arrangement (one is in the middle). The number of divisors is color-coded by hue. It starts with red, turns into yellow, green, blue, and violet.

Plate 2: Minimal Surfaces
Top left: two views of Henneberg's surface, first rendered with *Mathematica*, then with ray tracing.
Top right: Enneper's surface, rendered with ray tracing. Bottom: a curly surface.

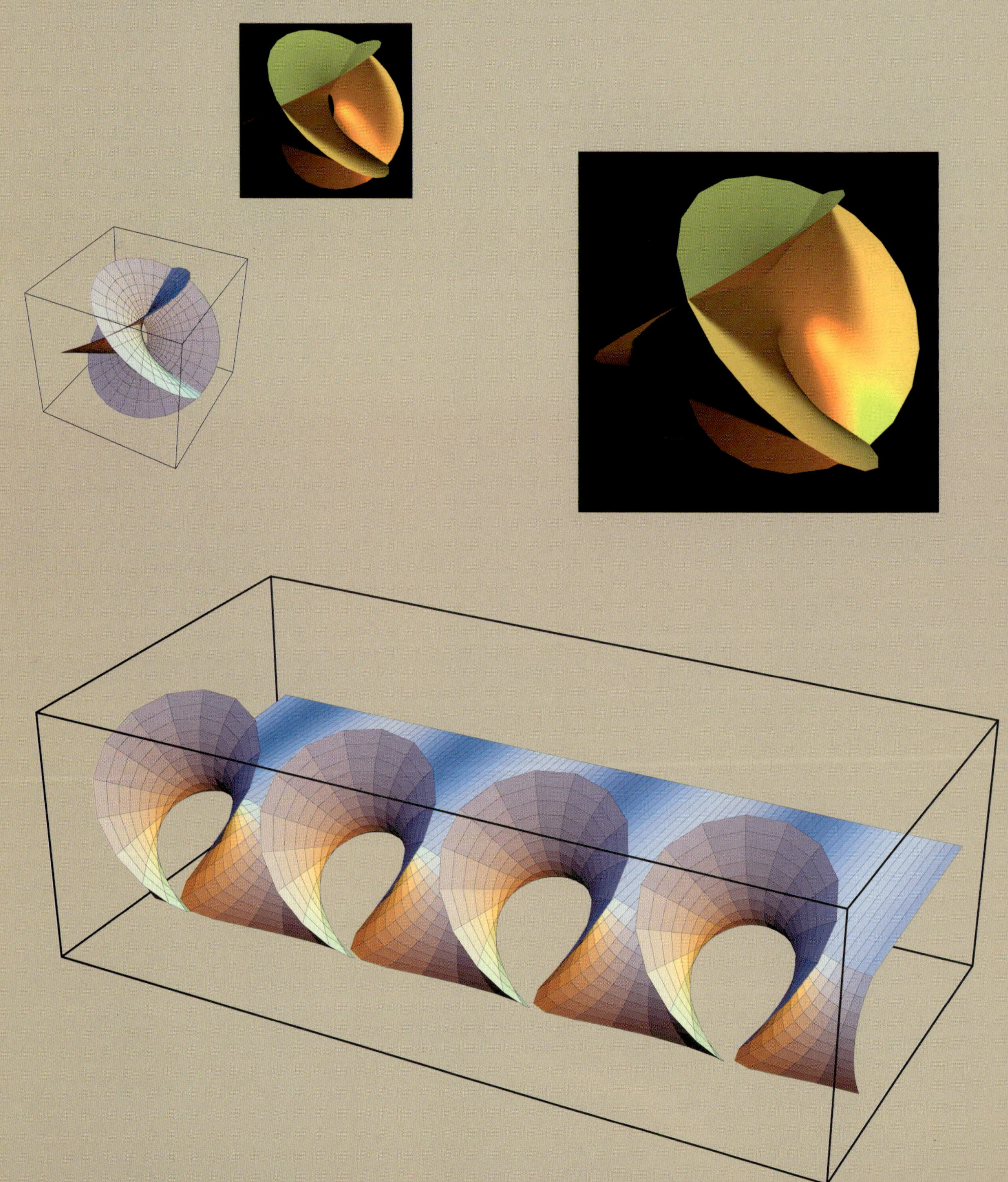

Plate 3: Uniform Polyhedra
The top row shows the great icosidodecahedron, rendered once with AVS, which adds simulated lighting, then with ray tracing, which also adds shadows. Middle row left: the small stellated dodecahedron; right: the great dodecahedron. Bottom row left: the great stellated dodecahedron; right: the great icosahedron.

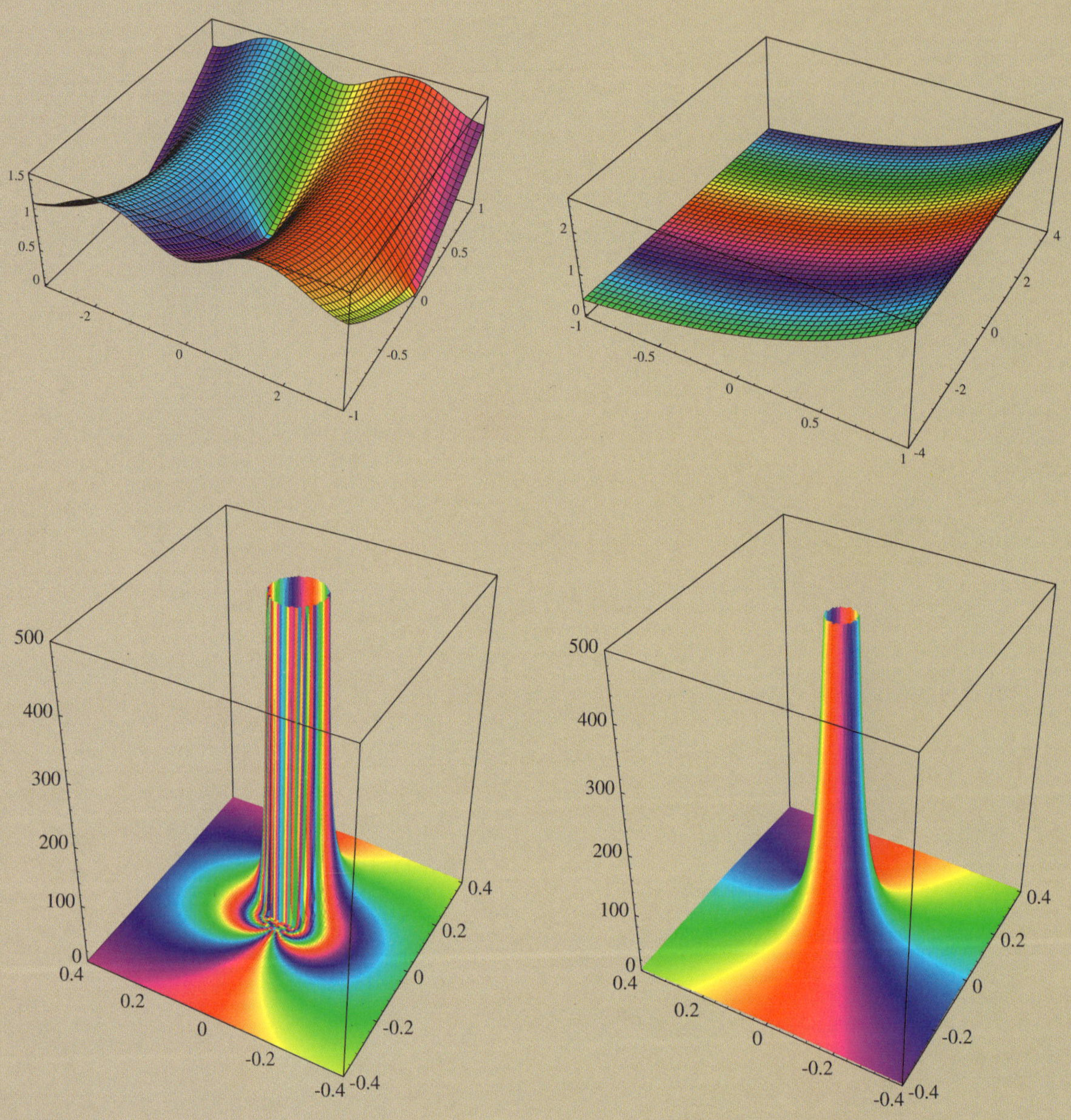

Plate 4: Complex-Valued Functions

Visualizations of complex-valued functions. The domain $\zeta = x + iy$ is represented in the x-y plane. The height of the surface (z-axis) is equal to the absolute value $|f(\zeta)|$. The argument or phase angle $\arg(f(\zeta))$ is color-coded by hue. Red is positive real. Upper left: $\sin(\zeta)$, upper right: e^ζ. Lower left: $e^{1/\zeta}$, which has an essential singularity at 0. Lower right: $1/\zeta^2$, which has a double pole at 0.

Plate 5: Sierpinski Sponge
A rendering of the third iteration of the fractal Sierpinski Sponge. The sponge has a metal-like surface (with specular highlights). The plane on which it rests has a grassy texture, achieved by random permutation of the surface normals (which influences the amount of reflected light). One light source is to the right of the eye point, another one is in the middle of the object.

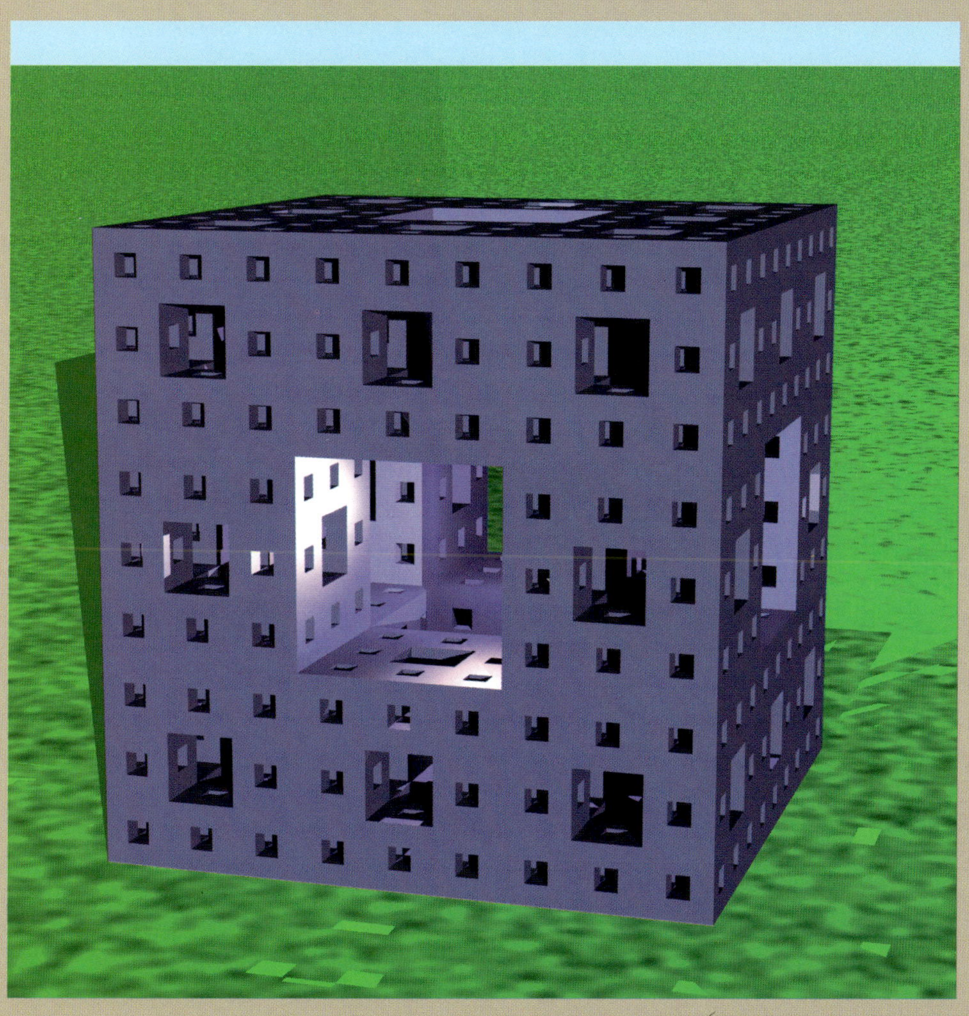

Plate 6: Stellated Icosahedra
Twelve stellated icosahedra. The solids are placed in front of a wall with wood-like texture. There are two point lights above to the left and right. The material of the solids shows specular reflections.

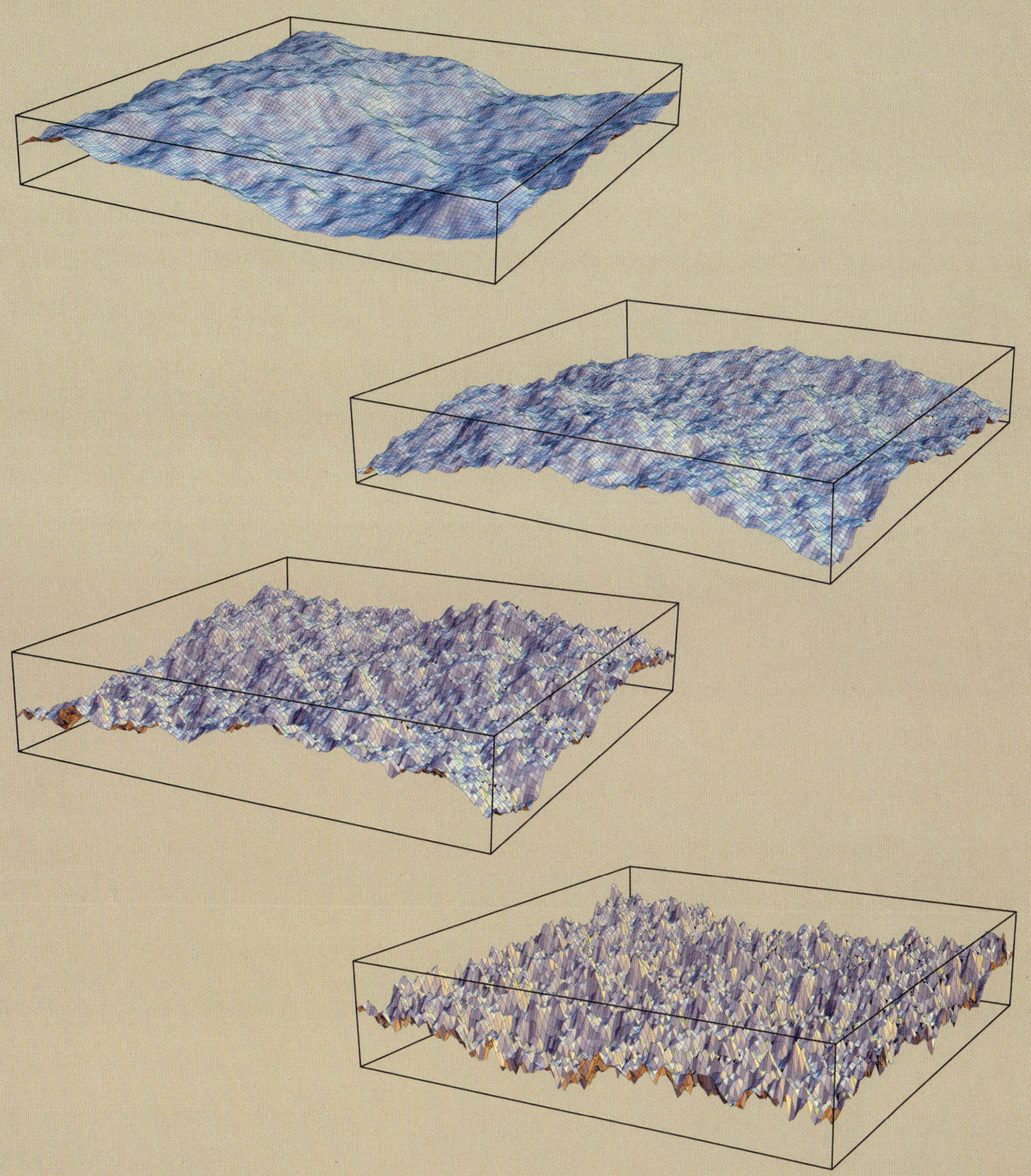

Plate 7: Fractal Mountains
A process called *fractional Brownian motion* (fBm) creates realistic-looking mountainous surfaces (this method is fully described in *The Science of Fractal Images* by Peitgen and Saupe). The fractal dimension varies from 2.2 to 2.8 from top to bottom.

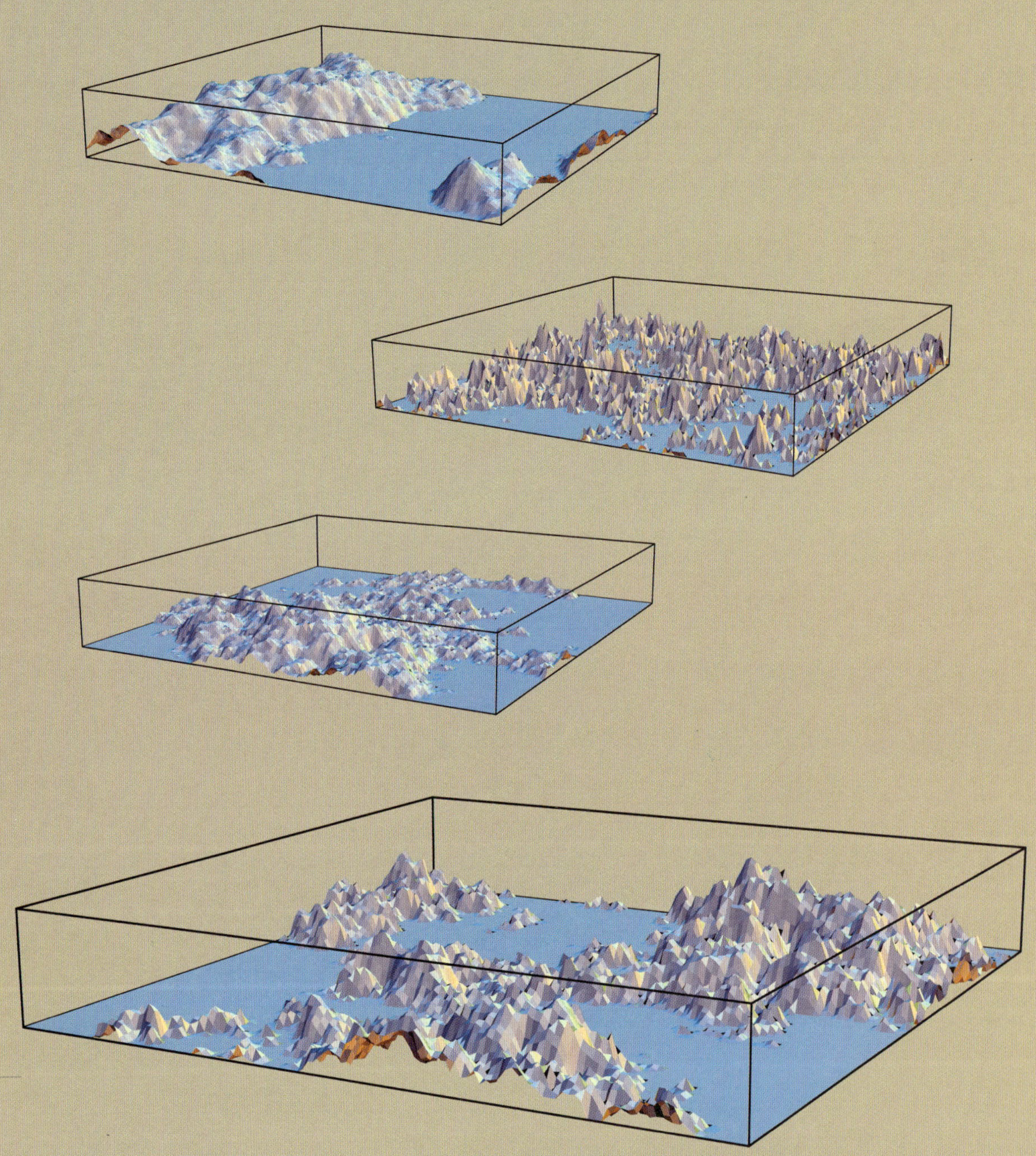

Plate 8: Fractal Beaches
The images from Plate 7, rendered in a different
way. All surface elements with a height below zero
are shown flat with sky blue color. This simple
change creates islands and lakes. The vertical
dimension has been fixed in all four pictures and the
mesh lines have been turned off.

```
          ]
        ];
          res
      ]

SetAttributes[FindOrAdd, HoldFirst]
FindOrAdd[rel_Symbol, keys_List, presets_List:{}, key_:None] :=
    Module[{ fields = Fields[rel], n = Length[keys], i, keyvals, keypos = 0,
                rest, restvals, prefields, prevals, qfields, tuple },
        keyvals = Array[Null, n];
        prefields = #[[1]]& /@ presets;
        prevals   = #[[2]]& /@ presets;
        For[i = 1, i <= n, i++, (* read keys[[i]] *)
                keyvals[[i]] = InputField[ keys[[i]] ];
                If[ keyvals[[i]] === EndOfFile, Return[EndOfFile] ];
        ];
        query = And @@ Thread[ Join[keys, prefields] == Join[keyvals, prevals] ];
        tuple = FindOne[ rel, query ];
        If[ tuple =!= $Failed,
            Print["Found: ", tuple]; Return[tuple] ];
        (* get rest of fields *)
        If[ key =!= None,
            keypos = Position[fields, key][[1,1]];
            fields = Drop[ fields, {keypos} ];
        ];
        qfields = Complement[ fields, prefields ]; (* query fields *)
        rest = Complement[ qfields, keys ]; n = Length[rest];
        restvals = Array[0, n];
        For[i = 1, i <= n, i++, (* read rest[[i]] *)
                restvals[[i]] = InputField[rest[[i]]];
                If[ restvals[[i]] === EndOfFile, Return[EndOfFile] ];
        ];
        tuple = fields /. Thread[ keys -> keyvals ] /.
                Thread[ rest -> restvals ] /. Thread[ prefields -> prevals ];
        If[ key =!= None,
            i = addWithKeyTo[ rel, tuple, key ];
            tuple = Insert[ tuple, i, keypos ];
          , addTupleTo[ rel, tuple ]
        ];
          tuple
      ]

Format[rel_relation] := R[ Fields[rel], Short[Tuples[rel], 1] ]

Protect[ Evaluate[protected] ]
Protect[ relation ] (* avoid saving Format, etc. *)

End[ ]

Protect[ Type, Prompt, Fields, Tuples, newRelation ]

EndPackage[ ]
```

Part 2
Applications

Chapter Six
Computations with Infinite Structures

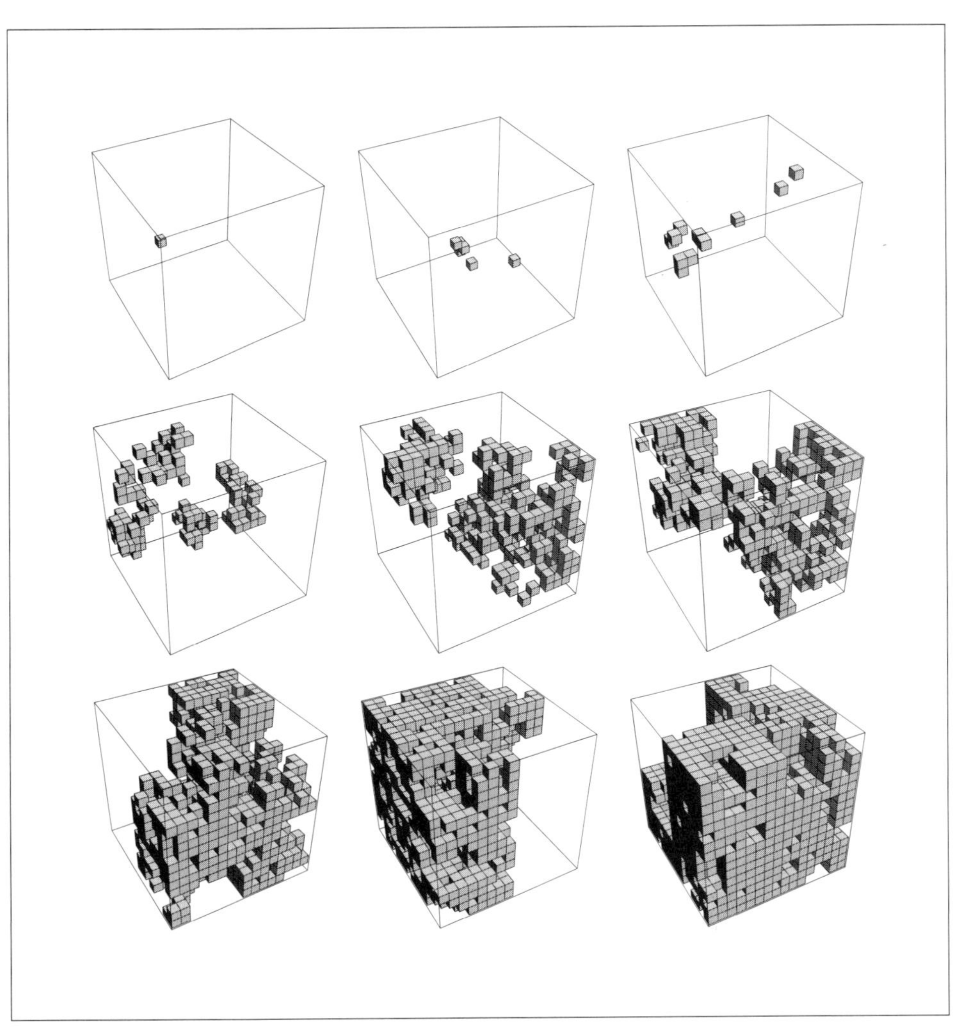

Looking at *infinite lists,* we find ways of computing with objects that do not fit inside any computer's storage all at once. We will define rules for the built-in list manipulation functions to deal with these infinite objects in the same way as with ordinary lists.

About the illustration overleaf:
Fractal cubes with different probabilities of recursively filling in the eight octants, from 0.1 (upper left) to 0.9 (lower right). They were produced with the command

```
Show[ GraphicsArray[
        Partition[ FractalCube[4, #]& /@ Range[0.1, 0.9, 0.1], 3],
        GraphicsSpacing->0]
]
```

using the package FractCube.m (see also Pictures.m).

6.1   Infinite Sets and Infinite Lists

It is, of course, impossible to deal with an *infinite* set explicitly on a finite computer. But a set whose elements can be generated one at a time, each in a finite number of steps—such a set is called *recursively enumerable*—can be considered in some sense as entirely "known," because any element can be produced on demand. The set of all perfect squares, for example, is recursively enumerable, because each element can be computed by the formula n^2, where $n = 1, 2, \ldots$. Such a formula describes a function of the integers; we can therefore define a recursively enumerable set as the *range* of a computable (or *recursive*) function.

Not all infinite sets can be generated in this way. A famous counterexample is the set of all real numbers. It is uncountable and, therefore, cannot be enumerable. There are also countable sets that are not recursively enumerable, for example the set of integers that cannot be expressed as the difference of two primes. Its elements cannot be found in a finite number of steps.

The same remarks made for sets (orderless collections of elements without repetitions) apply equally well to *streams,* or infinite lists. Like lists, streams are ordered and the same element can appear more than once. Being infinite, they have a first element, but no last one.

We can represent a stream in *Mathematica* by Stream[*first, rest*], where *first* is the first element and *rest* is an expression that will produce the rest of the stream when evaluated. The key point is that this expression is not evaluated immediately—only when subsequent elements of the stream are needed. This technique of evaluating an expression only when it is needed is termed *lazy evaluation.* Its implementation is trivial: we just give Stream the attribute HoldRest. The value produced by *rest* when it is evaluated is normally another stream. Thus we never reach the end—the stream is effectively infinite.

Clearly, we cannot perform on streams some of the operations that are possible on ordinary lists, such as sorting and taking the last element. Yet certain fundamental operations are available: taking the first element (First); removing the first element and returning the rest (Rest); and adding an element at the front (Prepend). From these many other useful operations can be constructed, as this chapter will show.

Good software design principles suggest that we do not work on Stream objects directly, but only through an interface that hides the implementation details. Here the interface consists of a *constructor* MakeStream[*first, rest*] and two *selectors* First[*stream*] and Rest[*stream*]. Should the implementation need to change, only the definitions of these three functions will be affected; all other code written on top of them can remain unchanged.

```
SetAttributes[Stream, HoldRest]

Stream/: First[Stream[ e_, r_ ]] := e
Stream/: Rest[Stream[ e_, r_ ]] := r

SetAttributes[MakeStream, HoldRest]

MakeStream[e_, s_] := Stream[e, s]
```

In our simple implementation, MakeStream[*first*, *rest*] simply turns into Stream[*first*, *rest*], but we will always use MakeStream when creating a stream. Thus an extension of Prepend for streams would be coded as

```
Stream/: Prepend[s_Stream, elem_] := MakeStream[elem, s] .
```

Note that we attach this definition to Stream, not to Prepend. This is the recommended practice for extending built-in functions to new argument types.

Our first example is a stream of ze-roes.

```
In[1]:= zeroes = MakeStream[0, zeroes]
Out[1]= Stream[0, zeroes]
```

According to its definition, the first element is 0.

```
In[2]:= First[zeroes]
Out[2]= 0
```

And the rest of it is equal to itself.

```
In[3]:= Rest[zeroes]
Out[3]= Stream[0, zeroes]
```

We can take the rest several times.

```
In[4]:= Nest[ Rest, zeroes, 1000 ]
Out[4]= Stream[0, zeroes]
```

As you can see, zeroes always regenerates itself. For less trivial examples (the previous one is far from useless, however) we usually need a function that we can call with suitable arguments to produce the rest of a stream. The following definition produces a stream of integers starting from n.

The first element of the stream of integers from n is, of course, n itself, and the rest is the stream of integers from $n + 1$.

```
In[5]:= IntegersFrom[n_Integer] :=
            MakeStream[n, IntegersFrom[n+1]]
```

Notice that the second element has not been evaluated.

```
In[6]:= integers = IntegersFrom[1]
Out[6]= Stream[1, IntegersFrom[1 + 1]]
```

If we take the rest of integers 99 times and then take the first el-ement, we should get the 100th ele-ment.

```
In[7]:= First[ Nest[Rest, integers, 99] ]
Out[7]= 100
```

6.2   Streams, Like Lists

We just saw that we can use First and Rest to extract the n-th element of a stream, but that is very inconvenient. We would like to treat streams as much as possible like ordinary lists. The technique for this is *overloading* the existing list-manipulation functions to work with streams. We have already done so with `First` and `Rest`.

Here is the definition to take the *n*-th element of a stream. The construction `Part[`*list*`, `*n*`]`, most often written in the form *list*`[[`*n*`]]`, takes the *n*-th element of a list. We define it to work with streams as well:

First, we take the rest $n - 1$ times, then take the first element of the result.

```
In[8]:= Stream/: Part[s_Stream, n_Integer?Positive] :=
                        First[Nest[Rest, s, n-1]]
```

Taking the 100th integer now becomes much easier.

```
In[9]:= integers[[100]]

Out[9]= 100
```

For ordinary lists, `Part[`*list*`, -`*n*`]` would take the *n*-th element from the end. This is not possible with streams, so we restricted the argument to be a positive integer.

Another basic operation is to take an initial segment of a stream (returned as a list). We can give rules for `Take[`*list*`, `*n*`]`, the function that does this for lists.

This is the special case, where no elements are taken.

```
In[10]:= Stream/: Take[s_Stream, 0] := {}
```

In the general case, we iterate `Rest[]` and then take the first elements of the iterated rests of the stream.

```
In[11]:= Stream/: Take[s_Stream, n_Integer?Positive] :=
                        First /@ NestList[Rest, s, n-1]
```

`Take[]` makes it convenient to look at a few elements of a stream to see whether things behave as intended.

```
In[12]:= Take[integers, 10]

Out[12]= {1, 2, 3, 4, 5, 6, 7, 8, 9, 10}
```

6.2.1 Mapping Functions onto Streams

One of the most useful concepts of functional programming is the mapping of functions over lists. Together with pure functions, it allows the elimination of unnecessary loops and auxiliary variables. In this example, we map the function "square the argument" over a list.

The cryptic notation `#^2&` is a pure function that squares its argument.

```
In[13]:= #^2& /@ {1, 2, 3, 4}

Out[13]= {1, 4, 9, 16}
```

Mapping can easily be defined for streams, using the ideas already discussed. To map a function *f* onto the stream `Stream[`*first*`, `*rest*`]`, we construct a new stream whose first element is `f[`*first*`]`, and whose rest is the (recursive) mapping of `f` to *rest*:

This definition is similar to the definition of map in LISP.

```
In[14]:= Stream/: Map[f_, s_Stream] :=
                        MakeStream[ f[First[s]], Map[f, Rest[s]] ]
```

We can now implement the example mentioned in the introduction—computing the infinite stream of all squares.

First we generate the stream of all integers, starting from one, then we square them.

```
In[15]:= squares = #^2& /@ IntegersFrom[1]
                             2
Out[15]= Stream[1, (#1  & ) /@

            Rest[Stream[1, IntegersFrom[1 + 1]]]]]
```

Here are the first twelve squares.

```
In[16]:= Take[squares, 12]
Out[16]= {1, 4, 9, 16, 25, 36, 49, 64, 81, 100, 121, 144}
```

Here is the stream of all prime numbers.

```
In[17]:= primes = Prime /@ IntegersFrom[1]
Out[17]= Stream[2, Prime /@

            Rest[Stream[1, IntegersFrom[1 + 1]]]]]
```

Again, the first twelve of them.

```
In[18]:= Take[primes, 12]
Out[18]= {2, 3, 5, 7, 11, 13, 17, 19, 23, 29, 31, 37}
```

Applying a function of several arguments to many streams in parallel is an extension of the notion of threading a function over lists. Elements of the argument lists are taken in parallel and the list of results is returned.

This example shows how Thread[ ] works with ordinary lists.

```
In[19]:= Thread[f[{a1, a2, a3}, {b1, b2, b3}]]
Out[19]= {f[a1, b1], f[a2, b2], f[a3, b3]}
```

Here is a definition for threading streams. (If you type this in, you have to unprotect Thread first. This definition cannot be stored with Stream, since the tag Stream is nested too deeply.)

```
In[20]:= Thread[f_[s__Stream]] :=
              MakeStream[ First /@ f[s],
                        Thread[Rest /@ f[s]] ]
```

This expression does not simplify, since Plus does not know how to add two streams.

```
In[21]:= squares + primes
                             2
Out[21]= Stream[1, (#1  & ) /@

            Rest[Stream[1, IntegersFrom[1 + 1]]]] +

          Stream[2, Prime /@ Rest[Stream[1, IntegersFrom[1 + 1]]]]
```

With Thread[ ] we can now add corresponding elements of the two streams squares and primes.

```
In[22]:= Thread[%]
Out[22]= Stream[3, Thread[Rest /@
                             2
          (Stream[1, (#1  & ) /@

              Rest[Stream[1, IntegersFrom[1 + 1]]]] +

            Stream[2, Prime /@

              Rest[Stream[1, IntegersFrom[1 + 1]]]])]]
```

Here are the first ten elements in this stream whose n-th element is the sum of the n-th prime number and n^2.

```
In[23]:= Take[%, 10]

Out[23]= {3, 7, 14, 23, 36, 49, 66, 83, 104, 129}
```

Output Formatting

As the examples grow more complex and interesting, voluminous output starts to fill our computer screen. In most cases, the expression that produces the rest of a stream is hard to read and should be rendered at most in an abbreviated way. Here is an output format for streams that makes them look like lists with ellipses at the end.

Short[*expr*, 0.5] gives a half-line abbreviated form of *expr*.

```
In[24]:= Format[ Stream[e_, s_] ] :=
            SequenceForm[ "{", e, ", ",
                Short[HoldForm[s], 0.5], "...}" ]
```

Now the complicated functions lurking in the rest of the stream are hidden.

```
In[25]:= squares
                     2
Out[25]= {1, (#1  & ) /@ <<1>>...}
```

6.3 Infinity at Last

Given a stream, we can define the substream that contains only those elements that satisfy a given predicate, using Select[*stream*, *pred*]. Select is predefined for lists; here is its definition for streams.

If the first element satisfies the predicates it becomes the first element of the filtered stream, otherwise we have to go on looking for a first element.

```
In[26]:= Stream /: Select[s_Stream, p_] :=
            If[ p[First[s]],
                MakeStream[ First[s], Select[Rest[s], p] ],
                Select[Rest[s], p] ]
```

This gives another way of generating a stream of all primes: we first generate all integers, then select only those that are prime.

```
In[27]:= otherprimes = Select[ IntegersFrom[1], PrimeQ ]

Out[27]= {2, Select[Rest[<<1>>], <<1>>]...}
```

The first ten primes.

```
In[28]:= Take[otherprimes, 10]

Out[28]= {2, 3, 5, 7, 11, 13, 17, 19, 23, 29}
```

So far, we have never encountered any problems with our infinite streams, since any initial piece of the desired answer could be computed from an initial piece of the input streams. This is still true for the preceding example, but now we see that we have to compute more and more elements from the input to find a given element of the output. To find the 10th prime, we had to generate all the integers up to 29.

What happens if at some point the given predicate is not true for any of the following elements in the stream? If we were to select all even numbers from the stream of primes above, using `Select[otherprimes, EvenQ]`, the computation would never finish—it would keep looking at more and more elements of `otherprimes` to find one that is even. If you try it out, at first everything looks fine, since it finds the first even prime right away and does not try to find the next one yet.

<table>
<tr><td>The first even prime is 2. The rest of the stream is not yet evaluated.</td><td>

```
In[29]:= Select[otherprimes, EvenQ]

Out[29]= {2, Select[Rest[<<1>>], <<1>>]...}
```
</td></tr>
</table>

If you take the rest of this stream, however, you will find yourself reaching for the abort key soon.

We have encountered an example of a *partial recursive* function that is not *total*. Such a function can be undefined for certain values of its argument. If it is given as a computer program, this means that the computation will never finish (without necessarily entering a simple infinite loop). With the primes above we can tell that it will never finish, but there is no algorithmic procedure for deciding this question for any given function. This problem is *undecidable*.

This can also happen with a predicate. We can define `MemberQ[`*stream*`, `*elem*`]`, which returns `True` if *elem* is a member of *stream*. Can it ever return `False`? The answer is no; it will continue searching the stream for an occurrence of *elem*, even if there is none.

<table>
<tr><td>If the first element is the one we asked for, we can immediately return `True`, otherwise we continue searching in the rest of the stream.</td><td>

```
In[30]:= Stream/: MemberQ[s_Stream, el_] :=
            If[ First[s] === el, True, MemberQ[Rest[s], el] ]
```
</td></tr>
<tr><td>The number 1999 is indeed prime. (Before trying out this example, you should set `$RecursionLimit` to `Infinity`.)</td><td>

```
In[31]:= MemberQ[primes, 1999]

Out[31]= True
```
</td></tr>
</table>

Had we asked for `MemberQ[primes, 3999]`, we would never have gotten an answer.

6.4 More Applications

We can easily write an infinite version of `Nest[ ]` that returns the stream of successive applications of a function to an initial element.

The first element is x, the rest is f
nested with initial value f[x].

```
In[32]:= Nest[f_, x_] := MakeStream[ x, Nest[f, f[x]] ]
```

(The built-in Nest requires a third argument that tells it how many times to nest the function. We don't need it here, since we want to nest infinitely often!)

6.4.1 Taylor Series

We use Nest[] to define the stream
of successive derivatives of an ex-
pression.

```
In[33]:= DStream[e_, x_] := Nest[ D[#, x]&, e]
```

Here are successive derivatives of
the logarithm function.

```
In[34]:= logderivs = DStream[ Log[x], x ]
Out[34]= {Log[x], Nest[D[#1, x] & , <<1>>]...}
```

This shows derivatives zero through
seven.

```
In[35]:= Take[%, 8]
```

$$\text{Out[35]= } \left\{ \text{Log}[x],\ \frac{1}{x},\ -x^{-2},\ \frac{2}{x^3},\ \frac{-6}{x^4},\ \frac{24}{x^5},\ \frac{-120}{x^6},\ \frac{720}{x^7} \right\}$$

If we divide this by a stream of factorials, we get the stream of successive terms in the Taylor series of Log[x].

Here is a definition of a stream of
factorials starting with 0.

```
In[36]:= FactorialStream[0]  :=
            MakeStream[ 1, FactorialStream[1] ]
```

This is the general case.

```
In[37]:= FactorialStream[n_] :=
            MakeStream[ n, n*#& /@ FactorialStream[n+1] ]
```

Here is the stream of successive
terms in the Taylor series for the log-
arithm according to the formula

```
In[38]:= taylor =
            Thread[logderivs*(1/#& /@ FactorialStream[0])]
Out[38]= {Log[x], Thread[Rest /@ (<<2>>)]...}
```

$$\frac{f^{(n)}(x)}{n!}\ .$$

```
In[39]:= Take[taylor, 8]
```

$$\text{Out[39]= } \left\{ \text{Log}[x],\ \frac{1}{x},\ \frac{-1}{2\,x^2},\ \frac{1}{3\,x^3},\ \frac{-1}{4\,x^4},\ \frac{1}{5\,x^5},\ \frac{-1}{6\,x^6},\ \frac{1}{7\,x^7} \right\}$$

Finally, we evaluate the Taylor se-
ries at $x = 1$ to generate the Taylor
coefficients of $\log(x)$ at $x = 1$.

```
In[40]:= coeffs = (# /. x -> 1)& /@ taylor
Out[40]= {0, TooBig...}
```

Here are the first ten coefficients.

```
In[41]:= Take[coeffs, 10]

             1      1      1    1      1    1      1    1
Out[41]= {0, 1, -(-), -, -(-), -, -(-), -, -(-), -}
             2      3      4    5      6    7      8    9
```

6.4.2 The Golden Ratio

With `Nest` we can also produce better and better approximations to numbers that can be described as the fixed point of a function. Here is a stream of approximations to `GoldenRatio`, computed from the continued-fraction expansion of `GoldenRatio`.

This stream contains successive iterations of the function $x \mapsto 1 + \frac{1}{x}$ applied to the initial value one.

```
In[42]:= goldenstream = Nest[1+1/#&, 1]

Out[42]= {1, Nest[1 + <<1>> & , <<1>>]...}
```

The numerators and denominators are the Fibonacci numbers.

```
In[43]:= Take[ goldenstream, 12 ]

                     3   5   8   13   21   34   55   89   144   233
Out[43]= {1, 2, -, -, -, --, --, --, --, --, ---, ---}
                     2   3   5   8    13   21   34   55   89    144
```

Here is the 100th term.

```
In[44]:= goldenstream[[100]]

             573147844013817084101
Out[44]= ─────────────────────────
             354224848179261915075
```

It is accurate to 41 decimal places.

```
In[45]:= N[ % - GoldenRatio, 50 ]

                          -42
Out[45]= 3.564158 10
```

6.5 Really Taking Off

Looking at the Taylor series example in the preceding section, we still see a lot of clumsy syntax that would not be necessary if we were working with lists instead of streams. The reason is that many functions, including all arithmetic ones, thread themselves automatically over lists—they have the attribute `Listable`.

All arithmetic operators and mathematical functions map themselves onto the elements of lists.

```
In[46]:= 2*{1, 2, 3, 4, 5}

Out[46]= {2, 4, 6, 8, 10}
```

If the argument is a stream we have to perform the mapping ourselves be writing the operator as a pure function and using `Map[ ]` for streams.

```
In[47]:= Map[ 2*#&, IntegersFrom[1] ]

Out[47]= {2, (2 #1 & ) /@ <<1>>...}
```

Finding a definition that makes streams behave like lists in this respect is a bit tricky, but here it is.

```
Stream/: f_Symbol[a___, b_Stream, c___] :=
    MakeStream[ f @@ FirstOrDup /@ {a, b, c}, f @@ RestOrDup  /@ {a, b, c} ] /;
        MemberQ[Attributes[f], Listable]

FirstOrDup[s_Stream] := First[s]
FirstOrDup[s_] := s
RestOrDup[s_Stream] := Rest[s]
RestOrDup[s_] := s
```

Listable functions threading over streams

This definition applies to functions `f` with the attribute `Listable` if at least one of its argument is a stream. If all arguments of `f` are streams, this code is similar to `Thread`. But listability also entails that arguments that are not lists (or streams in our case) be replicated. This is done by the auxiliary functions `FirstOrDup` and `RestOrDup`. The package Streams.m, reproduced at the end of this chapter, also contains definitions for application of rules to streams, and for the system functions `Union` and `Accumulate`. While we cannot sort a stream, `Union[`*stream*`]` will remove all duplicate elements. This happens be filtering out (using `Select[ ]`) all elements already encountered. The code for `Union[ ]` contains an interesting optimization.

```
Stream/: Union[s_Stream] :=
    With[{s1 = First[s]},
        MakeStream[s1, Select[Union[Rest[s]], # =!= s1&]]
    ]
```

The simplest form would be

$$\texttt{MakeStream[First[s], Select[Union[Rest[s]], \# =!= First[s]\&]]} .$$

Since the second argument of `MakeStream[ ]` is not evaluated, the term `First[s]` remains like this, with the value of s inserted, since it is a pattern variable. This expression would be evaluated over and over again later. Using `With[{s1 = First[s]}, `*body*`]` it is evaluated and then inserted everywhere in *body*. The same idea is also useful in `Accumulate[`f`, `s_0`, `*stream*`]`.

With this, we can write natural-looking code for streams. The Taylor coefficients from the last section can now be written as if the streams were lists:

We nest the differentiation, divide by the factorial numbers and evaluate at $x = 1$.

```
In[1]:= Nest[D[#, x]&, Log[x]] / FactorialStream[0] /. x->1

Out[1]= {0, TooBig ...}
```

```
In[2]:= Take[%, 10]
```

$$\texttt{Out[2]= } \{0,\ 1,\ -(\tfrac{1}{2}),\ \tfrac{1}{3},\ -(\tfrac{1}{4}),\ \tfrac{1}{5},\ -(\tfrac{1}{6}),\ \tfrac{1}{7},\ -(\tfrac{1}{8}),\ \tfrac{1}{9}\}$$

Accumulating the sum of more and more terms multiplied by $(y-1)^n$, we get successive orders in the Taylor series around the point $x = 1$.

```
In[3]:= Accumulate[Plus, %% (y - 1)^IntegersFrom[0]]

Out[3]= {0, (0 + <<1>> & ) /@ <<1>> ...}
```

Here is the series to order ten.

```
In[4]:= %[[11]]
```

$$Out[4] = -1 - \frac{(-1 + y)^2}{2} + \frac{(-1 + y)^3}{3} - \frac{(-1 + y)^4}{4} +$$

$$\frac{(-1 + y)^5}{5} - \frac{(-1 + y)^6}{6} + \frac{(-1 + y)^7}{7} - \frac{(-1 + y)^8}{8} +$$

$$\frac{(-1 + y)^9}{9} - \frac{(-1 + y)^{10}}{10} + y$$

The stream of squares of integers also benefits from these definitions.

```
In[5]:= Take[ IntegersFrom[1]^2, 20 ]

Out[5]= {1, 4, 9, 16, 25, 36, 49, 64, 81, 100, 121, 144,

         169, 196, 225, 256, 289, 324, 361, 400}
```

6.6   Conclusions and Further Reading

The idea of streams is from *Structure and Interpretation of Computer Programs* [1]. Comparing their implementation in SCHEME (a dialect of LISP) with ours highlights the differences between the evaluators of LISP and *Mathematica*. More on the computation of fixed points in *Mathematica* can be found in [28].

The origin of LISP and its connection with the theory of computation can be seen in McCarthy's article from 1960 [34]. (This paper also explains the origin of the names of the LISP functions `car` and `cdr`.) Decidability and infinity occupies much of the classic *Godel, Escher, Bach: An Eternal Golden Braid* [20]. An excellent source for the development of the theory of computation, a field pioneered by Turing's results on computability, is the collection of articles *The Universal Turing Machine: A Half Century Survey* [19]. Turing's original paper appeared in 1936 [44]. The uncountability of the set of real numbers was shown by G. Cantor in 1874 [7]. The precise definitions of the terms *partial recursive function* and *recursively enumerable set* can be found in any text on the theory of computation, for example in Brainerd/Landweber [5].

The floppy disk contains the package Streams.m with the definition of all the functions described here. The examples are contained in the notebook StreamExamples.ma.

6.7 The Complete Code of Streams.m

```
BeginPackage["Streams`"]

Streams::usage = "Functions for manipulating infinite streams. The following system
    functions know about streams: First, Rest, Prepend, Take, Part, Map, Select, Union,
    Thread, Accumulate, ReplaceAll, Nest, MemberQ.
    Listable functions thread over streams as they do over lists."

MakeStream::usage = "MakeStream[e, s] make a new stream by prepending
    the element e in front of the stream s."

First::usage = First::usage <> " First[stream] returns
    the first element of a stream."

Rest::usage = Rest::usage <> " Rest[stream] returns stream
    without its first element."

Prepend::usage = Prepend::usage <> " Prepend[stream, elem] prepends elem to stream."

Take::usage = Take::usage <> " Take[s, n] returns the
    first n elements of a stream s."

Part::usage = Take::usage <> " s[[i]] returns the i-th element of a stream s."

Map::usage = Map::usage <> " Map[f, s] returns a stream
    with f applied to each element of s."

Select::usage = Select::usage <> " Select[s, crit] returns
    a stream of only those elements of stream that satisfy crit."

Union::usage = Union::usage <> " Union[s] returns a stream
    with all duplicate elements removed."

Interleave::usage = "Interleave[s1, s2, ...] interleaves the streams si."

Thread::usage = Thread::usage <> " Thread[f[s1, s2, ...]]
    threads f over the streams si."

Accumulate::usage = Accumulate::usage <> " Accumulate[f, s]
    returns the stream {s1, f[s1, s2], f[s1, f[s2, s3]],...}."

Nest::usage = Nest::usage <> " Nest[f, x] returns the stream {x, f[x], f[f[x]], ...}."

ConstantStream::usage = "ConstantStream[c] generates an infinite stream of c's."

FactorialStream::usage = "FactorialStream[n] returns the stream
    {n, n(n+1), n(n+1)(n+2),...}."

IntegersFrom::usage = "IntegersFrom[n] returns the stream of integers starting at n."

Begin["`Private`"]

protected = Unprotect[Thread, Nest]

SetAttributes[Stream, HoldRest]

Stream/: First[Stream[ e_, _ ]] := e
Stream/: Rest[Stream[ _, r_ ]] := r
```

```mathematica
SetAttributes[MakeStream, HoldRest]     (* THIS IS WHY IT WORKS! *)

MakeStream[e_, s_] := Stream[e, s]

(* {s1, s2,..., sn} *)

Stream/: Take[s_Stream, 0] := {}
Stream/: Take[s_Stream, n_Integer?Positive] := First /@ NestList[Rest, s, n-1]

(* s[[i]] *)

Stream/: Part[s_Stream, n_Integer?Positive] := First[Nest[Rest, s, n-1]]
Stream/: Part[s_Stream, n1_, n__] := Part[Part[s, n1], n]

(* Prepend[s, e] *)

Stream/: Prepend[s_Stream, e_] := MakeStream[e, s]

(* f /@ s --> { f[s1], f[s2], ... } *)

Stream/: Map[f_, s_Stream] := MakeStream[ f[First[s]], Map[f, Rest[s]] ]

(* { si_1, si_2, ... }, with p[si_j] true *)

Stream/: Select[s_Stream, p_] :=
    If[ p[First[s]], MakeStream[ First[s], Select[Rest[s], p] ], Select[Rest[s], p] ]

(* {s11, s21,..., sn1, s12, s22,..., sn2, s13, ... } *)

Interleave[s1_Stream, s___Stream] := MakeStream[ First[s1], Interleave[s, Rest[s1]] ]

(* Union[s]: remove duplicate elements *)

Stream/: Union[s_Stream] :=
    With[{s1 = First[s]},
        MakeStream[s1, Select[Union[Rest[s]], # =!= s1&]]
    ]

(* MemberQ[s, e]. Will never return False, but may run forever *)

Stream/: MemberQ[s_Stream, el_] := If[ First[s] === el, True, MemberQ[Rest[s], el] ]

(* {f[s11, s21,..., sn1], f[s12, s22,..., sn2], ...} *)

Thread[f_[s__Stream]] := MakeStream[ First /@ f[s], Thread[Rest /@ f[s]]]

(* make streams threadable *)

FirstOrDup[s_Stream] := First[s]
FirstOrDup[s_] := s
RestOrDup[s_Stream] := Rest[s]
RestOrDup[s_] := s

Stream/: f_Symbol[a___, b_Stream, c___] :=
    MakeStream[ f @@ FirstOrDup /@ {a, b, c}, f @@ RestOrDup  /@ {a, b, c} ] /;
        MemberQ[Attributes[f], Listable]

(* Accumulate *)

Stream/: Accumulate[f_, s_Stream] :=
    With[{s1 = First[s]},
        MakeStream[ s1, f[s1, #]& /@ Accumulate[ f, Rest[s] ] ]
    ]
```

```
(* replacements, done elementwise *)

Stream/: s_Stream /.  rules_ := # /.  rules & /@ s
Stream/: s_Stream //. rules_ := # //. rules & /@ s

(* {x, f[x], f[f[x]], ...} *)

Nest[f_, x_] := MakeStream[ x, Nest[f, f[x]] ]

(* formats *)

Format[ Stream[e_, s_] ] := SequenceForm[ "{", e, ", ", Short[HoldForm[s], 0.5], " ...}" ]

(* some generators *)

ConstantStream[c_] := MakeStream[c, ConstantStream[c]]

FactorialStream[0]  := MakeStream[ 1, FactorialStream[1] ]
FactorialStream[n_] := MakeStream[ n, n FactorialStream[n+1] ]

IntegersFrom[n_Integer] := MakeStream[n, IntegersFrom[n+1]]

Protect[ Evaluate[protected] ]

End[ ]

Protect[ MakeStream, Stream, ConstantStream, FactorialStream, IntegersFrom ]

EndPackage[ ]
```

Chapter Seven
Fibonacci on the Fast Track

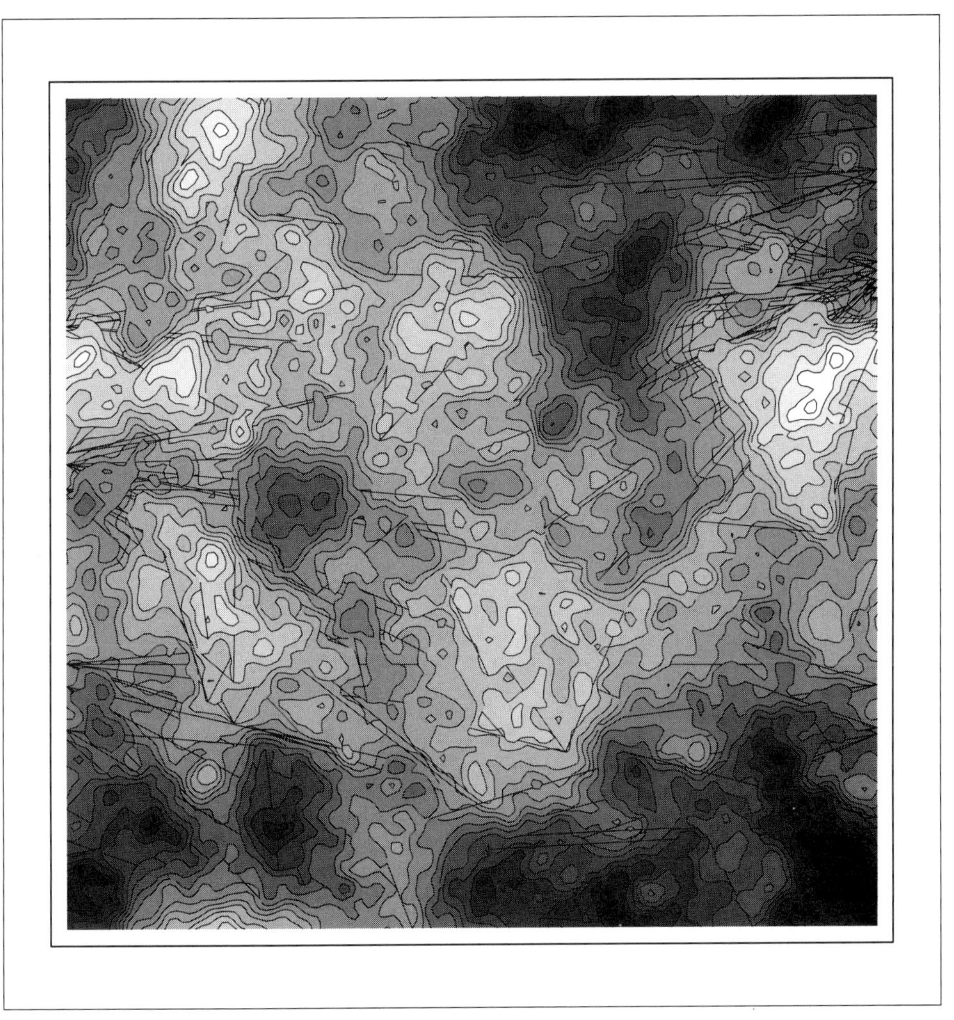

Ten progressively faster programs to compute Fibonacci numbers illustrate optimization techniques applicable to other problems as well.

About the illustration overleaf:
Contour lines of a randomly generated mountainous landscape. The topographical data was produced by a process called *fractional Brownian motion.* It is explained the book by Peitgen and Saupe [40] and was produced with the command (see Pictures.m):

```
Show[ ContourGraphics[fBm2D[128, 3 - 2.4, 64]],
     Contours -> 20, FrameTicks -> None ]
```

Other illustrations of such landscapes are shown on Plates 7 and 8.

7.1   Fibonacci Numbers

The *sequence of Fibonacci numbers* is easily described. The first two Fibonacci numbers f_1 and f_2 are both equal to 1. Each of the following numbers is equal to the sum of its two predecessors, $f_n = f_{n-1} + f_{n-2}$. These definitions can be programmed directly into *Mathematica*. They, as well as all other programs in this Chapter, are part of the package Fibonacci.m.

```
In[1]:= fiba[1] = fiba[2] = 1;

In[2]:= fiba[n_] := fiba[n-1] + fiba[n-2]

In[3]:= Table[ fiba[i], {i, 1, 10} ]
Out[3]= {1, 1, 2, 3, 5, 8, 13, 21, 34, 55}
```

This method of computing Fibonacci numbers is very inefficient. The number of recursive calls to compute fiba[*n*] is equal to f_n itself. As we will see, f_n grows exponentially with n.

7.1.1   Dynamic Programming

With *dynamic programming* we can improve performance. Each value found is immediately stored as a new rule for fibb.

```
In[4]:= fibb[1] = fibb[2] = 1;
```

Definitions are grouped right-to-left. The definition for fibb[n_] is itself a program to compute the value and assign it to fibb[n].

```
In[5]:= fibb[n_] := fibb[n] = fibb[n-1] + fibb[n-2]
```

The result is the same as before.

```
In[6]:= fibb[ 5 ]
Out[6]= 5
```

Now we can see that all previous values have been permanently stored.

```
In[7]:= ?fibb
Global`fibb

fibb[1] = 1

fibb[2] = 1

fibb[3] = 2

fibb[4] = 3

fibb[5] = 5

fibb[n_] := fibb[n] = fibb[n - 1] + fibb[n - 2]
```

The disadvantage of this method is that all intermediate Fibonacci numbers are stored as rules. Entering these rules into the system and subsequent lookup is slow and uses a lot of storage space.

7.1.2 Iterative Computation

The recursive definition given in the preceding section can be turned into an iterative one. We keep two values around, f_i and f_{i-1}. Their sum is equal to f_{i+1}. Now we no longer need f_{i-1}, and we can reuse f_i and f_{i-1} to hold f_{i+1} and f_i. With a parallel assignment this can be coded very clearly.

```
fibc[n_] :=
    Module[{fi = 1, fi1 = 0},
        Do[ {fi, fi1} = {fi + fi1, fi}, {n - 1} ];
        fi
    ]
```

Iteration

The local variables `fi` and `fi1` are initialized to f_1 and $f_0 = 0$, respectively. Therefore the loop has to be executed $n - 1$ times to compute f_n.

7.2 Complexity Measures

Complexity theory is concerned with the asymptotic behavior of algorithms. The complexity of an algorithm is the time (or number of operations) it takes to do the computation as a function of the *input size*. The *order of magnitude* is of primary concern. Thus an algorithm that takes n^2 steps for any input of size n is inferior to one that takes only $n \log n$ steps, for example.

Since f_n is of size n, the measure to use is n. The loop that computes `fibc[n]` is executed n times. In the i-th iteration two numbers of size i are added, which takes i steps. The total number of steps is therefore $\sum_{i=1}^{n} i \approx n^2$. The recursive method `fiba` needs f_n recursive calls. This is *exponential* in n. We can see this fast increase in the time quite easily in a picture. We measure the timings of `fiba[i]` for $i = 10, 11, \ldots, 20$ and then plot the results.

The timings are the first element of the result of the function `Timing[ ]`. To get numerical values, we set the symbol Second to one.

```
In[7]:= Table[ Timing[fiba[i]][[1]] /. Second->1,
            {i, 10, 20} ]

Out[7]= {0.0666667, 0.1, 0.183333, 0.266667, 0.466667,
         0.683333, 1.15, 1.8, 3.03333, 4.7, 7.65}
```

Such a picture does not prove exponential growth. All we can see is that the values grow rapidly.

```
In[8]:= ListPlot[ %, PlotRange->All ];
```

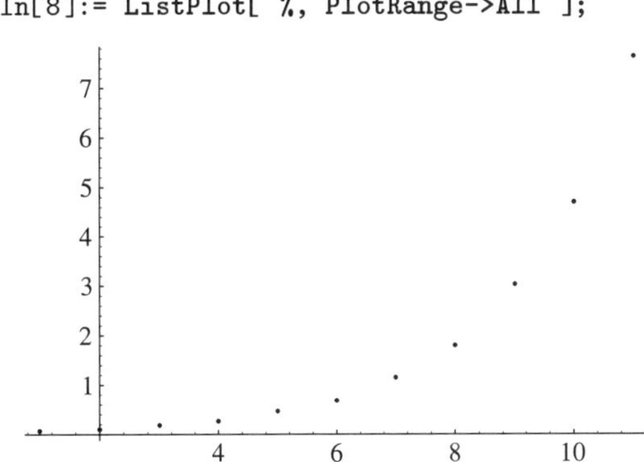

7.3　A Closed Formula

We would expect a much faster computation from a *closed formula,* rather than some kind of loop or even recursion. There is such a formula for Fibonacci numbers, but what mathematicians call a closed formula does not necessarily compute faster than a loop if large integers or multiprecision floating-point numbers are involved. Our example will illustrate this point quite clearly.

An iteration formula where each new term depends on the previous two can be solved for f_n in terms of n in closed form, that is, without explicit recursion or iteration. From the general iteration formula $f_{i+1} + b f_i + c f_{i-1} = 0$ we derive the *characteristic equation*

$$x^2 + bx + c = 0.$$

The solution of the iteration formula can then be expressed as

$$f_i = a_1 e_1^i + a_2 e_2^i, \qquad\qquad (7.3\text{--}1)$$

where e_1 and e_2 are the two solutions of the characteristic equation. For Fibonacci numbers we have $b = -1, c = -1$. In this computation we solve the characteristic equation and then assign the two solutions to the variables e1 and e2.

This gives us the two solutions of the quadratic equation.

```
In[1]:= Solve[ x^2 - x - 1 == 0, x ]

                1 - Sqrt[5]            1 + Sqrt[5]
Out[1]= {{x ->  -----------}, {x ->   -----------}}
                     2                      2
```

Now we assign the two solutions to the variables e1 and e2.

```
In[2]:= {e1, e2} = x /. %

               1 - Sqrt[5]   1 + Sqrt[5]
Out[2]= {----------------,  ----------------}
                 2                 2
```

The values of the constants a_1 and a_2 are computed from the initial conditions. In our case we have

$$f_1 = a_1 e_1 + a_2 e_2 = 1$$
$$f_2 = a_1 e_1^2 + a_2 e_2^2 = 1$$

(7.3–2)

Here is equation 7.3–2 in *Mathematica*. There is a unique solution.[1]

```
In[3]:= Solve[ {a1 e1 + a2 e2 == 1, a1 e1^2 + a2 e2^2 == 1},
               {a1, a2} ] // Simplify

                        1                    1
Out[3]= {{a1 -> -(---------), a2 -> ---------}}
                    Sqrt[5]            Sqrt[5]
```

We assign the (single) solution to the two unknowns.

```
In[4]:= {a1, a2} = {a1, a2} /. %[[1]]

                  1            1
Out[4]= {-(---------), ---------}
             Sqrt[5]     Sqrt[5]
```

This method is analogous to the way second-order linear differential equations are solved. In both cases we find two linearly independent solutions. The general solution is then a linear combination of the two. The coefficients of this linear combination are determined from the initial conditions.

Having determined all the unknowns, we can now define a closed form computation for f_n. Expanding out the formula $f_n = a_1 e_1^n + a_2 e_2^n$ gives us the n-th Fibonacci number.

```
In[5]:= fibd[n_] := Expand[ a1 e1^n + a2 e2^n ]
```

This method is very slow, since the expansion has to deal with the symbolic expression `Sqrt[5]`. At the end, all the square roots cancel and we get an integer value.

From this formula we can now compute the size of f_n. The base e_2 is smaller than 1 and will not contribute much to the result for large n. The size $\log f_n$ is therefore equal to

$$\log(a_1 e_1^n) = \log(a_1) + n \log(e_1) \approx n \log(e_1).$$

We see that the *size* of f_n is proportional to n.

While the preceding symbolic expansion was quite slow, this formula nevertheless leads to a fast way of computing f_n. For large n, the second term $a_2 e_2^n$ becomes very small. We can therefore compute the first term numerically and round to the nearest integer (we know that the result must be an integer). To get all the digits right, we must perform the computation with a precision equal to the size of f_n. We first find this size approximately by computing $a_1 e_1^n$ with machine precision. The number of digits needed is then the logarithm (to base 10) of the result. Now we do the computation a second time, with a precision a bit higher than the just-determined number of digits.

---

[1]Version 2.2 of *Mathematica* gives this result only with a prior option setting for `RowReduce[]` of `SetOptions[RowReduce, Method->CofactorExpansion]`. The way the solve functions are called is likely to be improved in future versions. More information can be found in the release notes.

```
fibe[n_] :=
   Module[{digits, approx},
      approx = N[a1 e1^n];
      digits = Ceiling[Log[10, approx]] + 10;
      approx = N[a1, digits] N[e1, digits]^n;
      Round[approx]
   ]
```

Numerical computation

An n-th power can be computed in just $\log n$ steps (more about this later). At each step two numbers of size n are multiplied together. The naive "schoolbook" method for multiplication of n digit numbers takes n^2 steps. The complexity of `fibe` is therefore $n^2 \log n$. This is worse than `fibc` with n^2 steps. For small n `fibe` is faster, but around $n = 3 \cdot 10^4$ `fibc` takes the lead. The exact value depends on the computer used, in our case a Sun SPARCstation 2.

Numbers can be multiplied faster than n^2. The *Karatsuba method,* for example, has a complexity of $n^{1.58}$. With such a method `fibe` is asymptotically faster. *Mathematica* uses such a method for integer multiplication, which we will now use. We denote the complexity of integer multiplication by $M(n)$.

7.4   The Matrix Method

The Fibonacci numbers appear as elements of the powers of the matrix

$$m = \begin{pmatrix} 1 & 1 \\ 1 & 0 \end{pmatrix} . \tag{7.4–1}$$

This follows from the fact that m has the same characteristic equation as the Fibonacci iteration. f_n is the top left entry in m^{n-1}, since

$$m^{n-1} = \begin{pmatrix} f_n & f_{n-1} \\ f_{n-1} & f_{n-2} \end{pmatrix} . \tag{7.4–2}$$

Here is the matrix m.

```
In[1]:= m = {{1, 1}, {1, 0}}

Out[1]= {{1, 1}, {1, 0}}
```

The ninth power has the Fibonacci numbers f_{10}, f_9, and f_8 as its elements.

```
In[2]:= MatrixPower[m, 9] // MatrixForm

Out[2]//MatrixForm= 55    34

                    34    21
```

The built-in `MatrixPower` computes high powers of integer matrices by the method of repeated squaring. Our own implementation as a program shows how it works. It also shows how powers can be computed in a logarithmic number of steps. The method looks at the binary representation of the exponent n. This corresponds to writing the exponent in the form

$$n = \sum_{i=0}^{k} n_i 2^i \, , \tag{7.4-3}$$

where n_i is 0 or 1. Therefore, m^n can be expressed as

$$m^n = \prod_{i=0}^{k} m^{n_i 2^i} = \prod_{i:\, n_i=1} m^{2^i} \, . \tag{7.4-4}$$

The powers m^{2^i} can be found by repeated squaring of m. For each 1 in the exponent, the current result is multiplied by the corresponding power of m. The bits of n are looked at from least significant to most significant. To see whether the least significant bit is 1, we test whether n is odd; n is then divided by 2, discarding any remainder. This corresponds to a right shift of the bits of n. The function `RepeatedSquaring` implements this method to raise a matrix m to the power n. The same method can be used to find powers of numbers or polynomials.

```
RepeatedSquaring[m_, n_] :=
    Module[{result = IdentityMatrix[2], nn = n, s = m},
        While[ nn > 1,
               If[OddQ[nn], result = result . s];
               s = s . s;
               nn = Quotient[nn, 2];
        ];
        result . s
    ]
```

Repeated squaring

We take the last multiplication of `result` by `s` outside of the loop, since we want to avoid squaring `s` once more. We would never use the result. This gives us our next method for computing Fibonacci numbers.

```
fibf[n_] := RepeatedSquaring[ {{1, 1}, {1, 0}}, n-1 ][[1,1]]
```

The matrix method for Fibonacci numbers (first version)

At each step we perform one or two matrix multiplications of 2 by 2 matrices. This takes eight integer multiplications. The entries of these matrices are certain Fibonacci numbers, doubling in size for each successive step. The number of steps is equal to the (binary) length of n. With $M(n) = n^2$, all the previous steps together take less time than the last one; therefore the complexity is n^2, the same as the iterative method `fibc`. It is about 10 times faster than `fibc`, however. (This shows that asymptotic complexity is not the only thing that matters.) The current implementation of integer multiplication has $M(n) = n^{1.58}$. This method is therefore *asymptotically faster* than the iterative computation.

Most of the work done is in the last matrix multiplication, of which we use only one element. We save about 50 percent of the total time by coding the last multiplication explicitly in terms of the elements of the matrices involved.

```
fibg[n_] :=
    Module[{result = IdentityMatrix[2], nn = n-1, s = {{1, 1}, {1, 0}}},
        While[ nn > 1,
               If[OddQ[nn], result = result . s];
               s = s . s;
               nn = Quotient[nn, 2];
        ];
        result[[1,1]] s[[1,1]] + result[[1,2]] s[[2,1]]
    ]
```

The matrix method for Fibonacci numbers (second version)

7.4.1   Accelerating the Matrix Method

Our matrices `result` and s are rather special. First, they are symmetric, meaning that $m_{12} = m_{21}$. Thus we need to compute only three of the four elements. Second, the entries are consecutive Fibonacci numbers. From any two of them we can compute the third one by addition or subtraction, since $f_n = f_{n-1} + f_{n-2}$ and $f_n = f_{n+1} - f_{n-1}$. Thus, from the four matrix elements we need to compute only two by multiplication. Denoting the elements of s by s_{ij} and those of `result` by r_{ij}, the result of squaring s, $t = s.s$ can be expressed as follows. (In the formulae we use different variable names for the result, to avoid confusion. In the programs we assign the result to the original variables, which are no longer needed.)

$$
\begin{aligned}
t_{11} &= s_{11}s_{11} + s_{12}s_{21} = s_{11}^2 + s_{12}^2 \\
t_{22} &= s_{21}s_{12} + s_{22}s_{22} = s_{12}^2 + s_{22}^2 \\
t_{12} &= t_{11} - t_{22} \\
t_{21} &= t_{12}
\end{aligned}
\qquad (7.4\text{--}5)
$$

and the result of $u = r.s$ is

$$
\begin{aligned}
u_{11} &= r_{11}s_{11} + r_{12}s_{21} = r_{11}s_{11} + r_{12}s_{12} \\
u_{22} &= r_{21}s_{12} + r_{22}s_{22} = r_{12}s_{12} + r_{22}s_{22} \\
u_{12} &= u_{11} - u_{22} \\
u_{21} &= u_{12}
\end{aligned}
\qquad (7.4\text{--}6)
$$

In each case, two of the four products needed are the same! We can therefore compute the matrix products with three instead of eight multiplications (two multiplications for the last step, since again we need only one element of the result). Instead of the matrices `s` and `result` we use variables for their elements. We do not store `s21` since it is the same as `s12`, which is also true for `r21` and `r12`.

```
fibh[n_] :=
    Module[{r11 = 1, r12 = 0, r22 = 1, s11 = 1, s12 = 1, s22 = 0, nn = n-1},
        While[ nn > 1,
            If[ OddQ[nn],
                {r11, r22} = r12 s12 + {r11 s11, r22 s22};
                r12 = r11 - r22
            ];
            {s11, s22} = s12^2 + {s11^2, s22^2};
            s12 = s11 - s22;
            nn = Quotient[nn, 2];
        ];
        r11 s11 + r12 s12
    ]
```

A Faster Matrix Method for Fibonacci numbers

7.4.2   Turning It Inside Out

It is not widely known that the power-tree method is not the best possible in this case. We can save half of the multiplications (in the worst case, where $n = 2^k - 1$, having all ones in the binary representation). Instead of working from least significant to most significant bit of n we start at the most significant bit. This is similar to Horner's rule for evaluating polynomials and corresponds to rewriting the formula for m^n in the following way:

$$m^n = \prod_{i=0}^{k} m^{n_i 2^i} = (\dots((m^{n_k})^2 m^{n_{k-1}})^2 \dots)^2 m^{n_0}. \tag{7.4--7}$$

In the loop we start out with $r = \mathbf{1}$, the identity matrix, and then repeatedly square r and, if $n_i = 1$, multiply it by m (m^{n_i} is equal to m or $\mathbf{1}$, since n_i is 1 or 0). The important observation is that the multiplication is always with the original m whose entries are all zeroes and ones. Multiplying with such a matrix involves only additions, no multiplications. For squaring r we use the same method, as in the previous case for squaring s. Computing $v = r.m$ gives

$$\begin{aligned}
v_{11} &= r_{11}m_{11} + r_{12}m_{21} = r_{11} + r_{12} \\
v_{22} &= r_{21}m_{12} + r_{22}m_{22} = r_{12} \\
v_{12} &= r_{11}m_{12} + r_{12}m_{22} = r_{11} \\
v_{21} &= v_{12}
\end{aligned} \tag{7.4--8}$$

The function `IntegerDigits[n, 2]` computes the digits of n in base 2 and returns them in a list. Since we know the number of iterations, we use a `Do` loop.

```
fibi[n_] :=
    Module[{r11 = 1, r12 = 0, r22 = 1, digits = IntegerDigits[n-1, 2], i},
        Do[ {r11, r22} = r12^2 + {r11^2, r22^2};
            r12 = r11 - r22;
            If[ digits[[i]] == 1,
                {r11, r12, r22} = {r11 + r12, r11, r12} ],
            {i, Length[digits]}
        ];
        r11
    ]
```

The use of Horner's rule

Again we should take the last iteration outside of the loop to avoid computing matrix entries that we never use. Trying to do this gives a surprising result.

7.4.3   Squeezing One More Out

Saving a multiplication in the last step is a bit trickier this time. If $n_0 = 0$, we need component t_{11} of the last squaring of $r.r = t$, which we can compute as $t_{11} = r_{11}^2 + r_{12}^2$, as we have seen. If $n_0 = 1$, we need $t_{11} + t_{12}$, where t is again the result of $r.r$. With some arithmetic, we find to our surprise that we can compute it with *one* multiplication!

$$
\begin{aligned}
t_{11} + t_{12} &= 2t_{11} - t_{22} \\
&= 2(r_{11}^2 + r_{12}^2) - (r_{12}^2 + r_{22}^2) \\
&= 2r_{11}^2 + r_{12}^2 - r_{22}^2 \\
&= 2r_{11}^2 + (r_{11} - r_{22})^2 - r_{22}^2 \\
&= 3r_{11}^2 - 2r_{11}r_{22} \\
&= r_{11}(3r_{11} - 2r_{22}) \\
&= r_{11}(r_{11} + 2r_{12})
\end{aligned}
\tag{7.4–9}
$$

This calculation shows that we can compute element v_{11} of $r.r.m$ in one multiplication as $r_{11}(r_{11} + 2r_{12})$. By shifting indices, we can compute v_{22} as $r_{12}(r_{12} + 2r_{22}) = r_{12}(r_{11} + r_{22})$. We use these two formulae in the loop for the case $n_i = 1$. In the case $n_i = 0$ we perform the same computations and then shift indices by one. Thus $t = r.r$ can be calculated as

$$
\begin{aligned}
t_{12} &= v_{22} = r_{12}(r_{11} + r_{22}) \\
t_{11} &= v_{12} = v_{11} - v_{22} = r_{11}(r_{11} + 2r_{12}) - t_{12} \\
t_{22} &= t_{11} - t_{12}
\end{aligned}
\tag{7.4–10}
$$

The one remaining problem is to find a formula for the last result in the case $n_0 = 0$ needing just one multiplication, as in the case $n_0 = 1$. The formula above for t_{11} needs two multiplications. For this, let us remember that the matrices r are powers of m. Let $r = m^k$ for some k. The operation $r.r.m$ corresponds to computing m^{2k+1}. The top left element of r is f_{k+1}, that of $r.r.m$

is f_{2k+2}. We have therefore derived a formula for computing f_{2n} in terms of f_n and f_{n-1} with one multiplication:

$$f_{2n} = f_n(f_n + 2f_{n-1}) = f_n(f_{n+1} + f_{n-1}). \tag{7.4--11}$$

This formula is not new. In the handbooks we find a similar formula for expressing Fibonacci numbers with an odd index:

$$f_{2n+1} = f_{n+1}(f_{n+1} + f_{n-1}) - (-1)^n. \tag{7.4--12}$$

Here, then, is the last function `fibj[]`, which uses just two multiplications in the loop, one outside.

```
fibj[n_] :=
    Module[{r11 = 1, r12 = 0, r22 = 1, digits = IntegerDigits[n-1, 2], i, t},
        Do[ If[ digits[[i]] == 1,
                {r11, r22} = {r11(r11 + 2r12), r12(r11 + r22)};
                r12 = r11 - r22
              , t = r12(r11 + r22);
                {r11, r12} = {r11(r11 + 2r12) - t, t};
                r22 = r11 - r12
              ],
            {i, Length[digits]-1}
          ];
        If[ digits[[-1]] == 1,
            r11(r11 + 2r12),
            r11(r11 + r22) - (-1)^((n-1)/2)
          ]
    ]
```

The fastest method

With this method we computed `fibj[10^7]`, the ten-millionth Fibonacci number, in 55 minutes on a SPARCstation 2. The result is approximately

$$1.12983437822539976032 \cdot 10^{2089876} \tag{7.4--13}$$

It has 2,089,877 digits.

7.5   Conclusions and Further Reading

All the methods derived in the last two sections have the same asymptotic complexity, $M(n)$. Nevertheless there is a huge difference in performance, about twenty-fold from `fibc` to `fibj`, which is, of course, not irrelevant. We have shown that we can compute f_n asymptotically as fast as we can multiply two n digit numbers. This is likely the best possible algorithm. For general recurrence relations, equation 7.3–1 is known as *Binet's equation;* it was discovered in 1843.

The improvements derived by finding better and better formulae for the matrix multiplications are, of course, specific to our problem, computing Fibonacci numbers. The derivation of asymptotically better algorithms by turning recursion into iteration and fast exponentiation can be used in other areas as well, such as numerical computation or computer algebra.

The history of Fibonacci numbers, originating in 1202 with Leonardo Pisano, or Leonardo Fibonacci (*Filius Bonaccii,* son of Bonaccio), and many of the formulae presented here, are given in Volume 1 of Knuth's book [21]. You will also find references to the various interesting uses of Fibonacci numbers there. Volume 2 of the same book discusses the binary methods for computing powers of polynomials, which are equally well suited for matrices, as we have seen [22]. The theory of linear iterative sequences is treated in any text on discrete mathematics or numerical analysis, for example [24] or [17].

The floppy disk contains the package Fibonacci.m with the definition of all the functions described here.

Chapter Eight
Fractal Curves

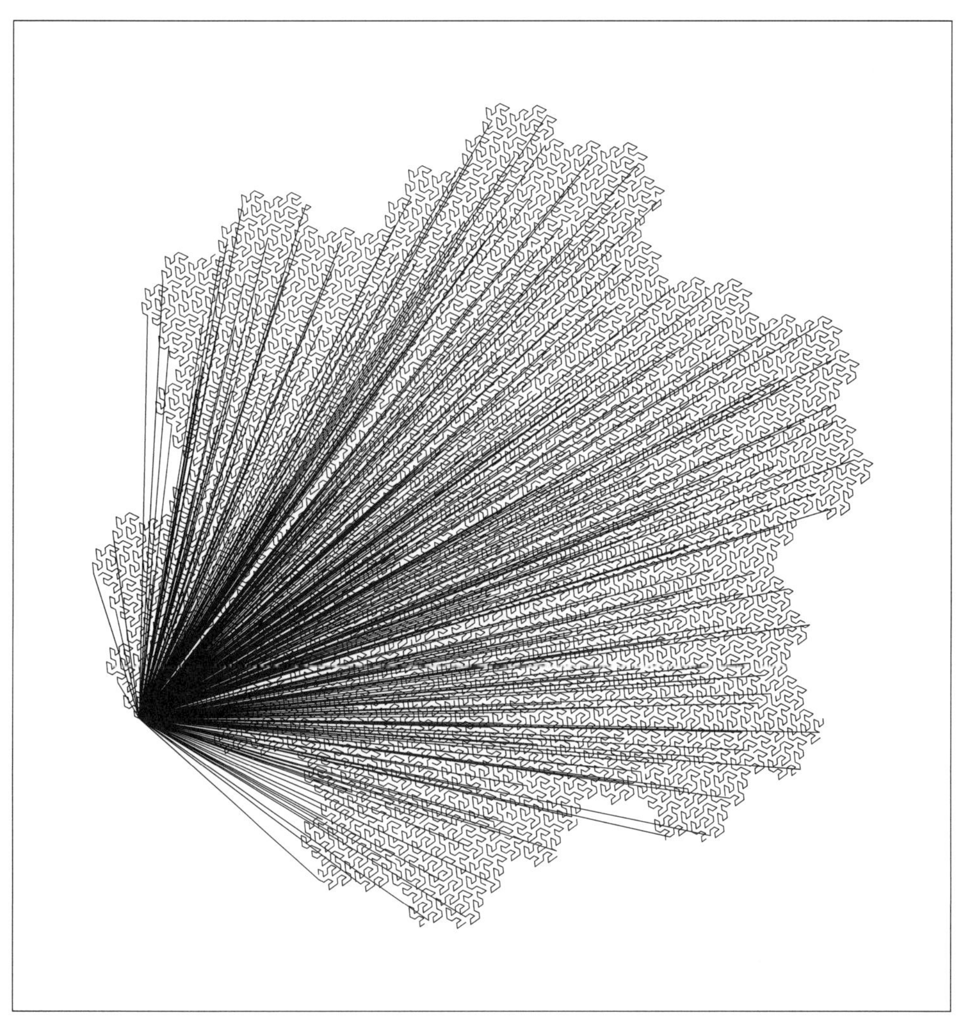

Fractal curves in the plane can be described by an iterative process of replacing lines by templates. We define our own language for describing such curves. Computing their graphical representation involves state transition systems. We will also develop a fast way of removing duplicates from a list.

8.1 Self-Similarity

Here is a description of a triangle, in a very natural "language."

```
In[1]:= triangle = Curve[ Move[1], Turn[120], Move[1],
                          Turn[120], Move[1] ];
```

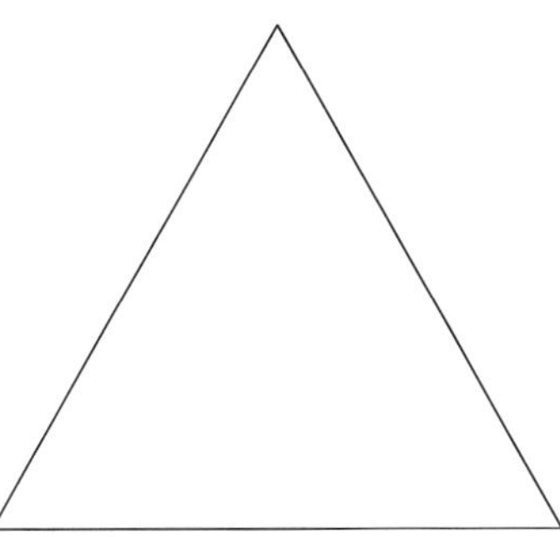

Replace each of its edges by a properly scaled and oriented copy of this template polygon.

```
In[3]:= template =
            Curve[ Move[1], Turn[-60], Move[1], Turn[120],
                   Move[1], Turn[-60], Move[1] ];
```

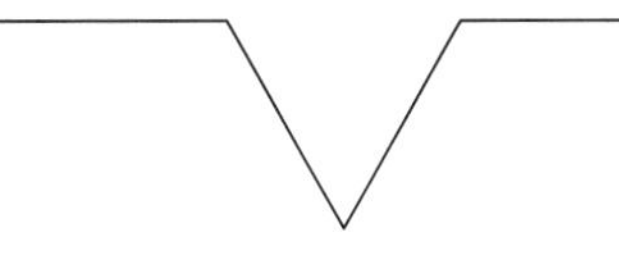

The result is this figure.

Now replace each edge of this polygon by a properly scaled and oriented copy of the template. And then again...

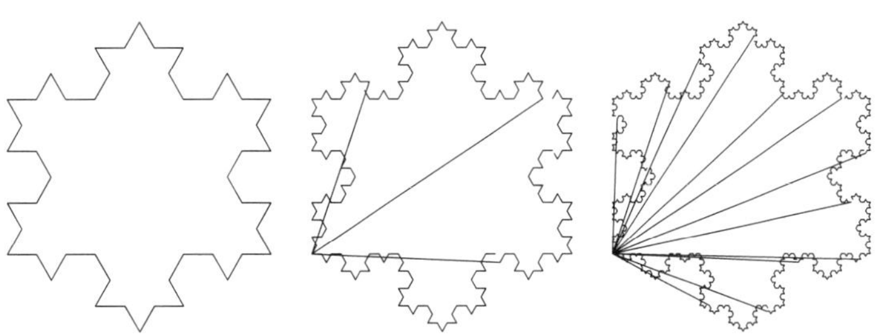

Repeat this process infinitely often. The result is the famous *Koch curve* or "snowflake," a *self-similar* curve in the plane: any part of it looks the same, no matter how much it is magnified.

8.2   A Language for Polygons

For our purposes the description of polygons in the plane in terms of coordinates is too low level and cumbersome. *Turtle graphics* comes in handy here. The turtle sits in the middle of a sheet of paper, facing east. You can give it two commands: *move* a certain distance and *turn* a certain amount. When moving, it leaves a black trail on the paper. A polygon can easily be described by these relative coordinates. In *Mathematica* we use the notation Move[*distance*] and Turn[*amount*]. A sequence of such commands describes a polygon, for which we use the name Curve, Polygon already being used. For example, the triangle with which we started the previous section was described by the curve

```
Curve[Move[1], Turn[120], Move[1], Turn[120], Move[1]] .
```

How do we now replace each edge of a polygon by a scaled and oriented copy of a template polygon? First, we make things easier by not worrying about the absolute size of our polygons. Scaling then simply means to multiply each distance measure in the template by the distance moved in the replaced move. Turns are not affected by scaling. This defines a multiplication of curves by numbers: A curve is multiplied by a number by multiplying each of its components according to the rules just described.

```
Curve /: r_ c_Curve := r # & /@ c

Move/: r_ Move[11_] := Move[r 11]
Turn/: r_ Turn[a_]  := Turn[a]
```

Scalar multiplication of curves

The replacement of a move in a curve by the template is now simply a substitution of each Move by a scaled copy of the template. The function Iterate[*curve, template*] performs this.

```
Iterate[c_Curve, template_Curve] := c /. Move[1_] :> 1 template
```

Here we see the curve that gave us the first iteration of the Koch curve.

```
In[7]:= Iterate[ triangle, template ] // InputForm

Out[7]//InputForm=
    Curve[Move[1], Turn[-60], Move[1], Turn[120], Move[1],
      Turn[-60], Move[1], Turn[120], Move[1], Turn[-60],
      Move[1], Turn[120], Move[1], Turn[-60], Move[1],
      Turn[120], Move[1], Turn[-60], Move[1], Turn[120],
      Move[1], Turn[-60], Move[1]]
```

We have already seen how this looks in graphic form. You might wonder why there are no nested curves as a result of the replacements performed: We have given `Curve` the attributes `Flat` and `OneIdentity`. A curve behaves like addition or multiplication in this respect; it is *associative*.

As with the built-in graphics objects `Graphics` and `Graphics3D`, we are usually not interested in their long-winded internal form but prefer an abbreviated output form. The definition to achieve this is straightforward:

```
Format[c_Curve] := "-Curve-" .
```

8.3   Displaying Fractals

It is neither necessary nor possible to iterate the replacement process an infinite number of times. The finite resolution of the output device (and of our eyes) allows us to display a finite approximation. The number of iterations necessary depends on the curve and the intended output device. The function `Fractal[`*initial*`, `*template*`, `*n*`]` performs n iterations. We have already defined the function that performs one iteration, so we simply nest this function n times:

```
Fractal[c_Curve, t_Curve, n_Integer?NonNegative] :=
        Nest[Iterate[#, t]&, c, n]
```

Computing the n-th approximation

The result of the function `Fractal` is a (usually rather complicated) curve, a sequence of move and turn commands. To display it, we have to convert it into a line that is understood by the built-in graphics code. A line is a list of vertex points given in Cartesian coordinates.

8.3.1   States and State Transitions

To compute the vertex coordinates, we use the model of *states* and *state transitions*. This model is used in many areas. The most interesting of these is a computer itself. At any given time, the computer is in a certain state, described by the contents of all of its registers and the memory. Each machine instruction changes that state. In this way a sequence of instructions performs a computation, whose result is some aspect of the last state achieved as a result of the stop instruction.

A formal description of this process uses a data type for the states and a data type for the possible instructions. The actions of the instructions are described by a function `NextState[`*state*`, `*cmd*`]`, which takes a state and an instruction into the state the system enters as a result of that instruction. The sequence of all states starting from an initial state s_0 under a sequence of instructions is then found by successively applying the next-state function. The function `FoldList[`*f*`, `s_0`, `*cmds*`]` performs this operation.

We can see what it does by using symbolic arguments.

```
In[8]:= FoldList[ f, s0, {cmd1, cmd2, cmd3} ]
Out[8]= {s0, f[s0, cmd1], f[f[s0, cmd1], cmd2],
     f[f[f[s0, cmd1], cmd2], cmd3]}
```

Back to turtle graphics. In cartesian coordinates the state of the turtle is described by its position $\{x, y\}$ and by its absolute orientation, measured in radians. The original orientation is taken to be 0 radians. We represent the state in the form $s[\{x, y\}, angle]$. The change of state under the move and turn commands is as follows:

- A *move* changes the position by adding a vector of the given length in the direction of the current orientation. The orientation does not change.

- A *turn* changes the orientation but not the position.

This leads to the following definition of NextState and the function CurveToStates for converting a sequence of moves and turns, namely, one of our curves, into a sequence of states:

```
deg = N[Degree]

NextState[s[xy_, a_], Move[l_, ___]] := s[xy + l {Cos[a], Sin[a]}, a]
NextState[s[xy_, a_], Turn[da_]]     := s[xy, a + da deg]

CurveToStates[c_Curve] :=
      FoldList[ NextState, s[{0.0,0.0}, 0.0], List @@ c ]
```

Updating the state

The pattern xy stands for the list $\{x, y\}$. Listability of the arithmetic functions makes this vector notation possible. The vector {Cos[a], Sin[a]} gives, of course, a unit vector in the required direction. deg is the numerical value of the conversion factor from degrees to radians. Remember that we chose the argument of Turn to be measured in degrees, since this is easier for humans.

8.3.2   Removing Duplicates in a List

We are almost there. The vertex coordinates of the line to be drawn are inside the state descriptors s. We simply extract them and wrap a Line around them. We remove any successive points with the same coordinates from the list. Many such points are generated because the turn commands do not change the position. An efficient way of removing these duplicates is not easy to find. There are many solutions. The important point is that all duplicates are removed in a *single* command, to avoid *repeated* changes in the length of the (long) lists involved. The function RemoveDuplicates is of general use.

```
RemoveDuplicates[l_List] :=
    Module[{pairs},
        pairs = Partition[ l, 2, 1 ];
        pairs = Select[ pairs, #[[1]] =!= #[[2]]& ];
        Append[ First /@ pairs, Last[l] ]
    ]

StatesToLine[st_List] :=
    Line[ RemoveDuplicates[st /. s[xy_, _] :> xy] ]
```

Converting a curve into a line

To see how it works, we set 1 to a symbolic example list.

```
In[9]:= l = {a, a, b, c, c, c, d, d};
```

The first line in the function creates a list of all pairs of successive elements of 1.

```
In[10]:= pairs = Partition[ l, 2, 1 ]
Out[10]= {{a, a}, {a, b}, {b, c}, {c, c}, {c, c}, {c, d},
   {d, d}}
```

Now we select all pairs that consist of two distinct elements.

```
In[11]:= pairs = Select[ pairs, #[[1]] =!= #[[2]]& ]
Out[11]= {{a, b}, {b, c}, {c, d}}
```

Finally, we retrieve the first elements of the pairs and add the last element of 1, which would otherwise be left out.

```
In[12]:= Append[ First /@ pairs, Last[l] ]
Out[12]= {a, b, c, d}
```

8.3.3 Curves as a Data Type

We can make these manipulations almost invisible to the user by treating our curves as a new type of graphics. We have to give rules to display them and to convert them into Graphics objects. This is done by overloading the existing commands Show, Display, and Graphics. The definitions should be stored with Curve. This is the preferred programming style: Curve is a new data type and we extend the domain of some built-in functions to this new type:

```
Curve/: Show[ c_Curve, opts___ ] :=
    Module[{thickness, gc},
        thickness = Thickness /. {opts};
        gc = Graphics[c];
        If[ NumberQ[thickness],
            gc[[1]] = Prepend[gc[[1]], Thickness[thickness]] ];
        Show[ gc, FilterOptions[Graphics, opts] ]
    ]

Curve/: Graphics[ c_Curve, opts___ ] :=
    Graphics[ {StatesToLine[CurveToStates[c]]}, opts, AspectRatio -> Automatic ]

Curve/: Display[ file_, c_Curve ] := Display[ file, Graphics[c] ]
```

Curves in graphics commands

Note that we allow an option Thickness->*val* in Show[ ]. This makes it easy to change the thickness of curves. Normally, Thickness is a graphics primitive, not an option. The preceding code inserts the primitive into the graphics object. FilterOptions[ ] removes this option and passes only valid options to Show[ ].

8.4 Variations on the Theme

The easy method to replace the line segments by the template curve only works if two conditions on the template are satisfied: The end point of the template must lie on the positive x-axis (due east from the start point) and the total amount of turn must be zero. Given *any* sequence of moves and turns in a template, these conditions can be achieved as follows:

- If the end point does not lie on the positive x-axis a counter rotation is prepended to the template. The amount of turn needed is the negative of the phase angle of the end point.

- If the total amount of turn does not add up to zero, a counter turn is appended.

Here is the definition of Fixup, implementing these measures. We can use the already defined function CurveToStates to compute the end point:

```
EndPoint[c_Curve] := Last[CurveToStates[c]] [[1]]

SetAttributes[ Fixup, Listable ]
Fixup[t_Curve] :=
    Module[{excess, tt = t, x, y},
        {x, y} = EndPoint[tt];
        rot = N[ArcTan[x, y]];
        If[ rot != 0.0, PrependTo[tt, Turn[-rot/deg]]];
        excess = Plus @@ First /@ Cases[tt, _Turn]; (* total rot. *)
        If[excess != 0.0, AppendTo[tt, Turn[-excess]] ];
        tt
    ]
```

Making a template canonical

For more complicated fractals we would like to have different kinds of edges in our curves. Each kind of edge will be replaced by a different template. We identify the edges by a number, given as a second argument in Move[*distance, kind*]. The function Iterate will then take a list of templates, replacing a Move[*d, n*] by the *n*-th template:

```
Iterate[c_Curve, template:{__Curve}] :=
    c /. Move[l_, n_:1] :> l template[[n]]
```

Different kinds of lines

Some of the other functions need to be updated, too, to allow for the extra argument inside `Move`. The details are in the complete package FractalCurves.m, reproduced at the end of this chapter.

For convenience, here are two generators for special curves. `Gon[n]` creates a *regular n-gon*. (A regular triangle being a 3-gon, a square a 4-gon, and so on. The 1-gon is simply a straight line.) `TurnCurve[{`a_1`, `a_2`, ..., `a_n`}]` gives a curve with unit moves in between turns of amounts $a_1, a_2, \ldots, a_n$:

```
Gon[n_] := TurnCurve[ Table[Turn[360.0/n], {n-1}]]

TurnCurve[ l_List ] :=
    Apply[ Curve,
           Transpose[{Table[Move[1], {Length[l]}],
                      Turn /@ l}],
           {0, 1}] ~Append~ Move[1]
```

Frequently used curves

8.5 The Fractal Dimension

If you scale an ordinary curve by a linear factor of $1/r$, you need r scaled copies to reconstruct the original. If you scale a square by a linear factor of $1/r$, you need r^2 pieces. The volume of a solid scales as r^3. The exponent in these formulae is the *dimension* of the geometrical figure in question. Now consider the Koch curve: The replacement template is made up of 4 pieces, each 1/3 the length of the original. The exponent in this case is equal to $\log_3 4 \approx 1.26186$. Therefore we say that it has (fractal) dimension $\log_3 4$.

The hexagonal fractal curve described next is made out of seven pieces, each of size $1/\sqrt{7}$. Its dimension is therefore $\log_{\sqrt{7}} 7 = 2$. It is a *space-filling* curve, having the same dimension as the plane.

In simple cases we can find the fractal dimension from our representation of the template curves. The scaling r is equal to the distance of the end point of the template from the origin. The number of pieces is equal to the length of the template, the total amount of movement:

```
CurveLength[c_Curve] := Plus @@ First /@ Cases[c, _Move]

FractalDimension[t_Curve] :=
    With[{xy = EndPoint[t]},
        N[ Log[Sqrt[xy.xy], CurveLength[t]] ]
    ]
```

The fractal dimension

This simple computation does not work if there are several different templates.

8.6 Examples

The title graphic of this chapter is the "flowsnake," a space-filling fractal curve derived from the hexagonal tiling of the plane. It uses two kinds of edges that are replaced according to the following templates. The two are mirror images of each other, so we can still compute their fractal dimension in the simple way outlined in the preceding section. The initial curve is simply a straight line. The title picture is the fifth iteration. It consists of 16807 line segments.

```
ht1 = Curve[Move[1, 2], Turn[-60], Move[1, 1], Move[1, 1], Turn[-120], Move[1,1 ],
        Turn[-60], Move[1, 2], Turn[120], Move[1, 2], Turn[60], Move[1, 1]]
ht2 = Reverse[ht1] /. {Turn[a_] :> Turn[-a], Move[d_, i_] :> Move[d, 3-i]}

Flowsnake[n_] := Fractal[ Curve[Move[1, 1]], {ht1, ht2}, n ]
```

Templates for the flowsnake

These pictures show the first four iterations of the flowsnake.

```
In[1]:= Show[ GraphicsArray[
                  Table[ Graphics[Flowsnake[n]], {n, 0, 3} ]
           ] ];
```

The curve with this replacement template shows *double points:* vertices that touch another part of itself.

```
In[2]:= temp = TurnCurve[{-90, 90, 90, -90}];
```

Here is its dimension.

```
In[3]:= FractalDimension[ temp ]

Out[3]= 1.46497
```

Here are the first four iterations.

```
In[4]:= Show[ GraphicsArray[
                 Table[Graphics[Fractal[ Gon[4], temp, i ]],
                       {i, 0, 3}] ]
           ];
```

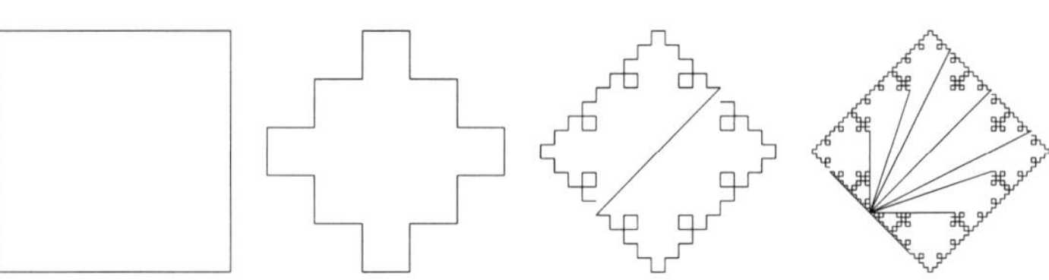

The *Hilbert curve* is another space-filling fractal curve. The first two templates are rotations and reflections of each other. The third one is a simple move that is not subdivided further. This is achieved by replacing it by itself.

```
hx = Curve[ Turn[-90], Move[1, 2], Move[1, 3], Turn[90], Move[1, 1], Move[1, 3],
            Move[1, 1], Turn[90], Move[1, 3], Move[1, 2], Turn[-90] ]
hy = Curve[ Turn[90], Move[1, 1], Move[1, 3], Turn[-90], Move[1, 2], Move[1, 3],
            Move[1, 2], Turn[-90], Move[1, 3], Move[1, 1], Turn[90] ]
hz = Curve[ Move[1, 3] ]

Hilbert[n_] := Fractal[ hx, {hx, hy, hz}, n ]
```

The Hilbert curve

Here are the first four iterations. They look a bit different from the usual representation. The move and turn commands used here are not general enough to represent it in the standard form. Only the finite approximations look different. In the limit we get the same fractal curve.

```
In[5]:= Show[ GraphicsArray[
              Table[ Graphics[Hilbert[n]], {n, 1, 4} ]
            ] ];
```

At right are another two well-known fractal curves, the *Koch island* (dimension 1.55105), and the *square Sierpinski curve* (dimension 1.77124).

```
In[6]:= Show[ GraphicsArray[{
              Graphics[Fractal[Gon[4],
                  TurnCurve[{90,-90,-90,0,0,90,90,-90}], 3]],
              Graphics[Fractal[Gon[4],
                  TurnCurve[{0,90,90,90,90,0}], 3]] }]
            ];
```

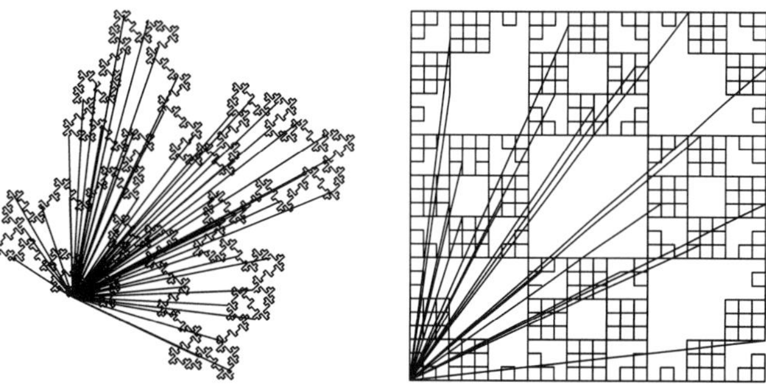

One motivation for the development of the theory of fractals was the study of natural phenomena, for example a coast line. At any scale it looks similar, with large and small bays and peninsulas. Coast lines or mountain ranges have been successfully modeled by fractals. Some randomness in the replacement templates gives them a more natural appearance. We can produce crude pictures of this kind by choosing simple templates with gentle turns and varying amounts of move. Here is an example.

We choose gentle turns and arbitrary variations of the arc lengths.

```
In[7]:= coast =
           Curve[
             Move[1], Turn[25], Move[1], Turn[30], Move[1.5],
             Turn[-30], Move[0.8], Turn[-35], Move[0.5]
           ];
```

Here is the fifth iteration.

```
In[8]:= Show[ Fractal[ Curve[Move[1]], coast, 5 ],
              AspectRatio -> Automatic ];
```

8.7   Conclusions and Further Reading

The classic references to fractals of all kinds are Mandelbrot's books [32, 33]. A more general variant of our curves, called *0L-systems,* is described in Peitgen and Saupe's *The Science of Fractal Images* [40]. This book contains a lot of material about computer-generated fractals, including a description of the methods used in some Hollywood movies.

One of their programs for generating fractal surfaces has been implemented in *Mathematica.* It is included on the floppy disk (package fBm.m). Some sample sceneries generated by this program are reproduced in Plates 7 and 8. The flowsnake curve is due to W. Gosper and is described in [13]. The original reference to the Koch curve is [47].

A collection of stunning photographs of (mostly) natural fractals and some computer-generated ones is found in a beautiful book by McGuire [35]. From it we take the idea for the last picture, a fractal curve attributed to Mandelbrot. It has the same outline as the snowflake curve. Its dimension is 1.86178, so it is not quite space-filling, despite appearances.

Turtle graphics was introduced in the programming language LOGO. It was later used in the UCSD-Pascal system on the Apple II.

The floppy disk contains the package FractalCurves.m with the definition of all the functions described. The notebook FractalExamples.ma contains the code for reproducing the examples.

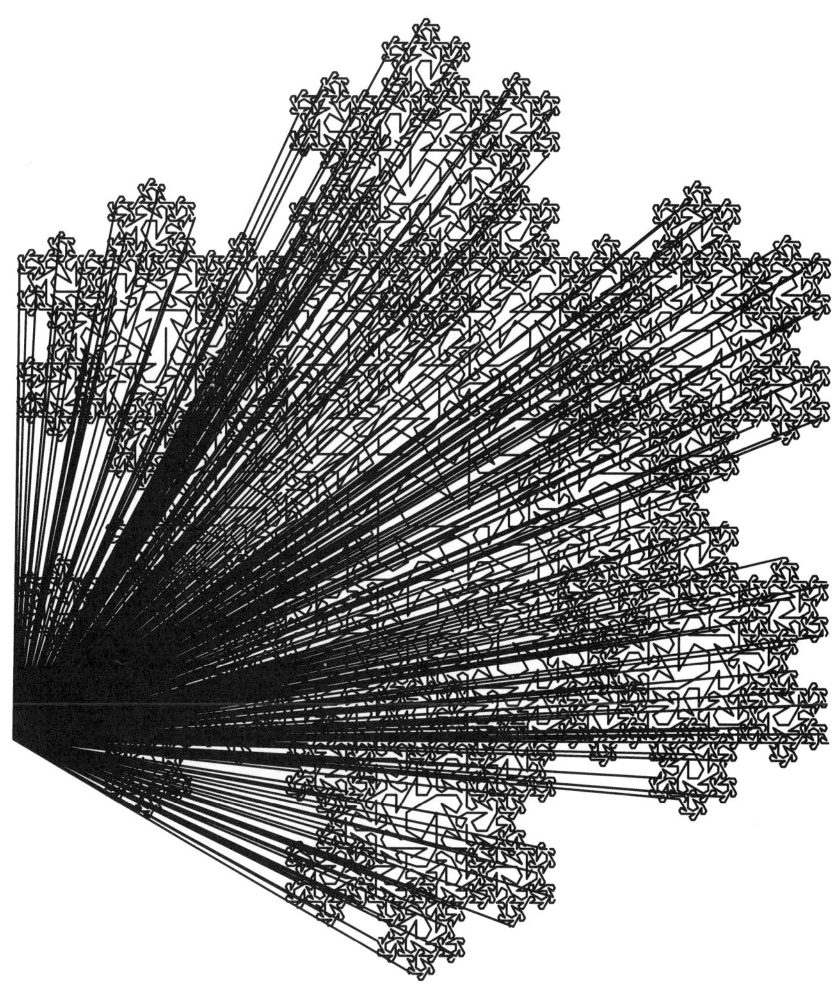

8.8 The Complete Code of FractalCurves.m

```
BeginPackage["FractalCurves`", "Utilities`FilterOptions`"]

Fractal::usage = "Fractal[curve, template, n] gives the n-th iterate of replacing all
    Move[]'s in curve by template. Fractal[curve, {template1,..}, n] replaces each
    Move[l, i] by template_i. Results can be converted to Graphics[] or shown
    directly with Show[]."

Iterate::usage = "Iterate[ curve, template ] replaces each Move[]
    in curve by a scaled copy of template."

Turn::usage = "Turn[a] represents a turn by the angle a, measured in degrees."

Move::usage = "Move[l] represents a move by l in the current direction."

Curve::usage = "Curve[ cmds... ] describes a sequence of moves and turns."

Gon::usage = "Gon[n_] gives a regular n-gon as a curve."

TurnCurve::usage = "TurnCurve[{a1,...,an}] gives the curve
    Curve[Move[1], Turn[a1], Move[1], Turn[a2],.., Turn[an], Move[1]]."

FractalDimension::usage = "FractalDimension[template] computes the
    fractal dimension of a fractal constructed with template."

Begin["`Private`"]

SetAttributes[Curve, {Flat, OneIdentity}]

(* scaling of curves *)

Curve /: r_ c_Curve := r # & /@ c

(* one iteration *)

Iterate[c_Curve, template_Curve] := c /. Move[l_] :> l template

Iterate[c_Curve, template:{__Curve}] := c /. Move[l_, n_:1] :> l template[[n]]

(* multiplication *)

Move/: l_ Move[ll_, args___] := Move[l ll, args]
Turn/: l_ Turn[a_]   := Turn[a]

(* n-th approximation to a fractal *)

Fractal[c_Curve, t_, n_Integer?NonNegative] :=
    With[{tt = Fixup[t]},
        Nest[Iterate[#, tt]&, c, n]
    ]

(* convert to absolute coordinates *)

deg = N[Degree]

NextState[s[xy_, a_], Move[l_, ___]] := s[xy + l {Cos[a], Sin[a]}, a]
NextState[s[xy_, a_], Turn[da_]]      := s[xy, a + da deg]

CurveToStates[c_Curve] := FoldList[ NextState, s[{0.0,0.0}, 0.0], List @@ c ]
```

```
RemoveDuplicates[l_List] :=
    Module[{pairs},
        pairs = Partition[ l, 2, 1 ];
        pairs = Select[ pairs, #[[1]] =!= #[[2]]& ];
        Append[ First /@ pairs, Last[l] ]
    ]

StatesToLine[st_List] := Line[ RemoveDuplicates[st /. s[xy_, _] :> xy] ]

(* display *)

Curve/: Show[ c_Curve, opts___ ] :=
    Module[{thickness, gc},
        thickness = Thickness /. {opts};
        gc = Graphics[c];
        If[ NumberQ[thickness],
            gc[[1]] = Prepend[gc[[1]], Thickness[thickness]]
          ];
        Show[ gc, FilterOptions[Graphics, opts] ]
    ]

Curve/: Graphics[ c_Curve, opts___ ] :=
    Graphics[ {StatesToLine[CurveToStates[c]]}, opts, AspectRatio -> Automatic ]

Curve/: Display[ file_, c_Curve ] := Display[ file, Graphics[c] ]

(* aux *)

EndPoint[c_Curve] := Last[CurveToStates[c]] [[1]]

CurveLength[c_Curve] := Plus @@ First /@ Cases[c, _Move]

(* the fractal dimension *)

FractalDimension[t_Curve] :=
    With[{xy = EndPoint[t]},
        N[ Log[Sqrt[xy.xy], CurveLength[t]] ]
    ]

(* make sure, angles add up to 0 and total orientation is 0 *)

SetAttributes[ Fixup, Listable ]
Fixup[t_Curve] :=
    Module[{excess, tt = t, x, y},
        {x, y} = EndPoint[tt];
        rot = N[ArcTan[x, y]];
        If[ rot != 0.0, PrependTo[tt, Turn[-rot/deg]]];
        excess = Plus @@ First /@ Cases[tt, _Turn]; (* total rot. *)
        If[excess != 0.0, AppendTo[tt, Turn[-excess]] ];
        tt
    ]

(* Format *)

Format[c_Curve] := "-Curve-"

(* convenience generators for initial curves *)

(* regular n gon *)
```

```
Gon[n_] := TurnCurve[ Table[360.0/n, {n-1}] ]

(* make curve with equal moves between turns *)

TurnCurve[ l_List ] :=
    Apply[ Curve,
           Transpose[{Table[Move[1], {Length[l]}],
                       Turn /@ l}],
           {0, 1} ] ~Append~ Move[1]

End[ ]

Protect[ Curve, Move, Turn, Fractal, Iterate, TurnCurve, Gon ]

EndPackage[ ]
```

Chapter Nine
Minimal Surfaces

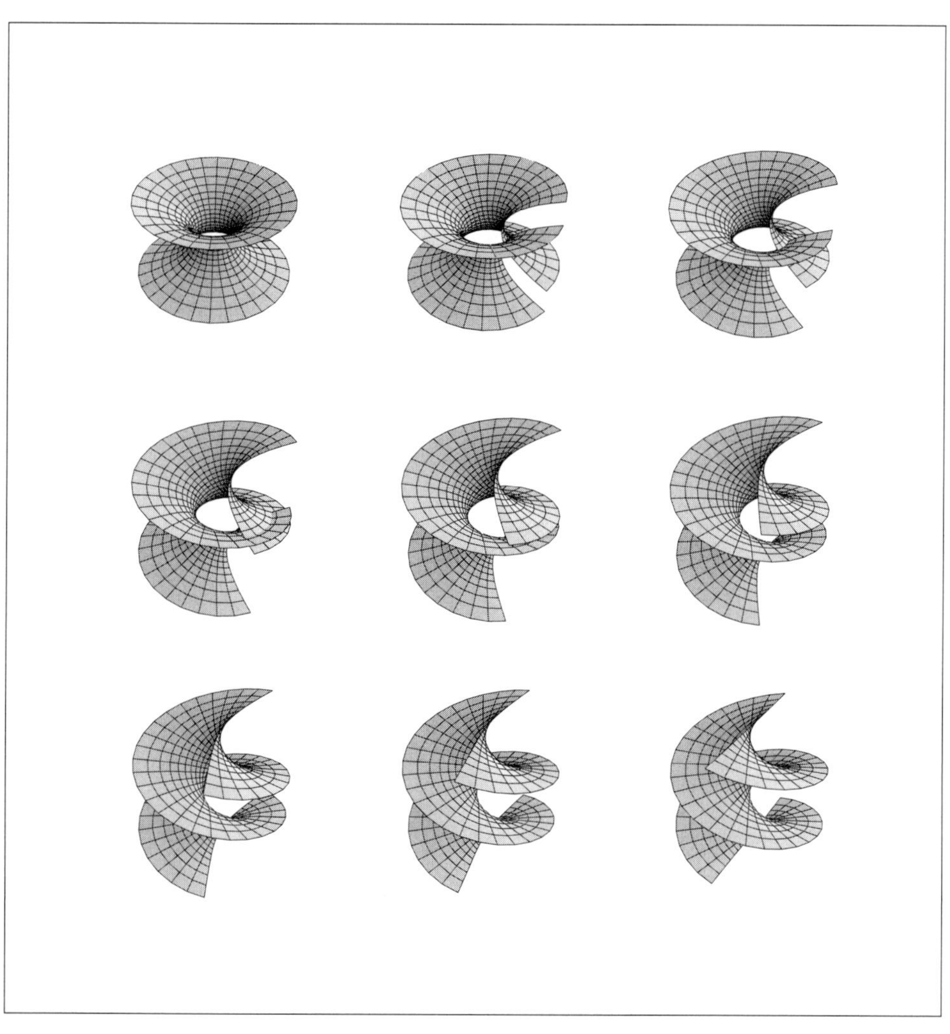

Minimal surfaces have a strange beauty and an amazing variety of shapes. We investigate a simple formula that can produce an infinite number of minimal surfaces. This topic provides a good illustration of the combination of symbolic computation, numerical evaluation, and graphics within *Mathematica*. Symbolic simplification is necessary to find analytic continuations of the multivalued functions used to describe the surfaces. We use animation to show the continuous transformation of one surface into another.

About the illustration overleaf:
The transformation of a catenoid (top left) into a helicoid (lower right); see Section 9.4.

9.1　A Formula for Minimal Surfaces

Minimal surfaces are surfaces with mean curvature zero. One way to define the mean curvature of a surface at a point P is in terms of the normal curvature. The normal curvature at P in a given direction is the reciprocal of the radius of curvature of a curve in the surface passing through P in that direction. The mean curvature is the average of the maximum and minimum values of this quantity over all directions at P. A trivial example of a minimal surface is a plane. At any point, the curvature is zero in all directions, so the mean curvature is identically zero. If the mean curvature of a surface is not identically zero, the maximum curvature must be positive and the minimum curvature must be negative for the average to be zero. This gives the minimal surfaces their twisted look. Another characterization is as a surface of minimal area. A film of soap spanned by a closed wireframe assumes minimal area among all surfaces with this wireframe as boundary.

9.1.1　A Parametric Representation

Surfaces in space can be described parametrically by giving the three coordinates x_1, x_2, and x_3 as differentiable functions of two parameters u and v

$$\mathbf{x}(u, v) = \{x_1(u, v), x_2(u, v), x_3(u, v)\}. \tag{9.1–1}$$

The parameterization is called *regular* if the following two tangent vectors

$$\begin{aligned}
\frac{\partial \mathbf{x}}{\partial u} &= \left\{\frac{\partial x_1}{\partial u}, \frac{\partial x_2}{\partial u}, \frac{\partial x_3}{\partial u}\right\} \\
\frac{\partial \mathbf{x}}{\partial v} &= \left\{\frac{\partial x_1}{\partial v}, \frac{\partial x_2}{\partial v}, \frac{\partial x_3}{\partial v}\right\}
\end{aligned} \tag{9.1–2}$$

are linearly independent at each point (u, v). In this case, they span the *tangent plane* of the surface at the point.

It is convenient to introduce the complex parameter $\zeta = u + iv$ and describe the surface in terms of complex-valued coordinate functions. From the real coordinate functions $\{x_1, x_2, x_3\}$, define complex functions $\{\phi_1, \phi_2, \phi_3\}$ by

$$\phi_k(\zeta) = \frac{\partial x_k}{\partial u} - i\frac{\partial x_k}{\partial v}, \quad k = 1, 2, 3. \tag{9.1–3}$$

The parameterization is called *isothermal* if

$$\phi_1(\zeta)^2 + \phi_2(\zeta)^2 + \phi_3(\zeta)^2 = 0. \tag{9.1–4}$$

It turns out that a surface described by an isothermal parameterization is minimal if and only if the coordinate functions $x_k(u, v)$ are harmonic. A function $h(u, v)$ is called *harmonic* if

$$\frac{\partial^2 h}{\partial u^2} + \frac{\partial^2 h}{\partial v^2} = 0. \tag{9.1–5}$$

This is equivalent to the condition that the complex functions $\phi_k(\zeta)$ are analytic.

We can therefore describe a minimal surface by a triple $\{\phi_1, \phi_2, \phi_3\}$ of analytic functions such that $\sum \phi_k{}^2 = 0$. The real parameterization is given by

$$x_k = \operatorname{Re} \int \phi_k(\zeta) d\zeta. \tag{9.1-6}$$

An important result of complex analysis lets us obtain any such triple of analytic functions from an analytic function f and a meromorphic function g as

$$\begin{aligned}
\phi_1(\zeta) &= f(1 - g^2) \\
\phi_2(\zeta) &= if(1 + g^2) \\
\phi_3(\zeta) &= 2fg
\end{aligned} \tag{9.1-7}$$

with the additional condition that f must have a zero of order at least $2m$ at each pole of g of order m. A meromorphic function is a function that is analytic except at isolated poles. The function $1/\zeta$, for example, is analytic except for $\zeta = 0$, where it has a simple pole.

From this we obtain a minimal surface as

$$\operatorname{Re} \int \{ f(1 - g^2), if(1 + g^2), 2fg \} d\zeta \tag{9.1-8}$$

This formula is called the *Enneper–Weierstrass parameterization*. The surface is regular if f has only the zeroes at the poles of g described in the preceding paragraph. A simple way to make sure that this condition is satisfied is to not include any singularities of g or zeroes of f in the chosen parameter range.

9.1.2 Enneper's Surface

We now show some examples of minimal surfaces constructed with the Enneper-Weierstrass parameterization. See also the articles by Dickson [11] and Ogawa [37].

One of the simplest choices for f and g is $f(\zeta) = 1$, $g(\zeta) = \zeta$.

```
In[1]:= f = 1; g = z;
```

We can use *Mathematica's* indefinite integration command to obtain the three coordinate functions.

```
In[2]:= s = Integrate[{f(1-g^2), I f(1+g^2), 2 f g}, z]

                 3
                z            I  3   2
Out[2]= {z  -  --,  I z  +  - z , z }
                3            3
```

To draw a picture of this surface, we have to express the complex parameter z in terms of two real ones, then take the real part of the coordinate functions and simplify. One way is to write z in cartesian coordinates as z = u + I v.

```
In[3]:= s /. z -> u + I v

                     3
             (u + I v)                        I         3
Out[3]= {u - --------- + I v,  I (u + I v) +  - (u + I v) ,
                 3                            3

                  2
         (u + I v) }
```

Re takes the real part of an expression and `ComplexExpand` simplifies it.

```
In[4]:= ComplexExpand[ Re[%] ]
```

$$Out[4]= \{u + \frac{-u^3 + 3\,u\,v^2}{3}, \ -v + \frac{-3\,u^2\,v + v^3}{3}, \ u^2 - v^2\}$$

Now we can use the built-in command for parametric plotting. Here is a picture of Enneper's surface for $-2.5 \le u \le 2.5$, $-2.5 \le v \le 2.5$.

```
In[5]:= ParametricPlot3D[ Evaluate[%],
        {u, -2.5, 2.5}, {v, -2.5, 2.5}
     ];
```

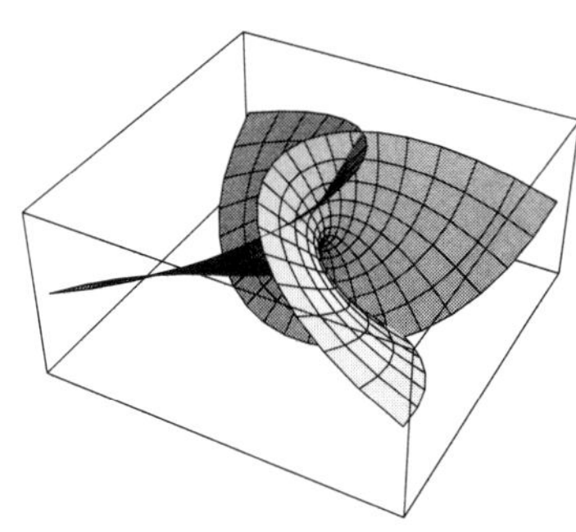

Alternatively, we can express `z` in polar coordinates as $z = re_{i\phi}$.

```
In[6]:= s /. z -> r Exp[I phi];
```

Again we take the real part and simplify the result.

```
In[7]:= ComplexExpand[Re[%]]
```

$$Out[7]= \{r\,Cos[phi] - \frac{r^3\,Cos[3\,phi]}{3},$$

$$-(r\,Sin[phi]) - \frac{r^3\,Sin[3\,phi]}{3}, \ r^2\,Cos[2\,phi]\}$$

This gives us a different view of the same surface. See also Plate 2 for a ray-traced rendering.

```
In[8]:= ParametricPlot3D[ Evaluate[%],
        {r, 0, 2.5}, {phi, 0, 2Pi},
        PlotPoints->{12, 36}, PlotRange->All
     ];
```

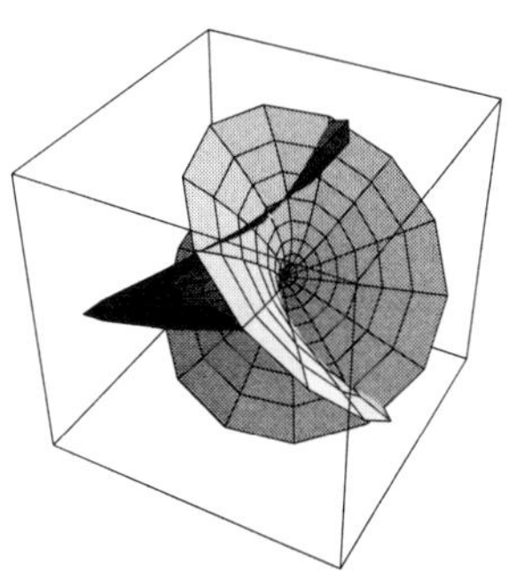

9.2   A Package for Plotting Minimal Surfaces

Our approach to plotting the minimal surface given by the two functions f and g is similar to the example (complex maps) presented in Chapter 1 of "Programming in *Mathematica*." We will develop a package with two commands, CartesianSurface and PolarSurface. Each takes the expressions for f and g and the name of the variable z as inputs, along with the ranges for the two parameters. Let us look at the auxiliary functions needed for this.

The function Parameterize[f, g, z] performs the integration. Since the integration is not guaranteed to succeed we should examine the result and issue an error message in case it fails.

```
Parameterize[f_, g_, z_] :=                  (* the integration *)
    Module[{list},
        list = Integrate[{f - f g^2, I(f + f g^2), 2 f g}, z];
        If[ !FreeQ[list, Integrate],
                Message[MinimalSurfaces::noint, list]; Abort[] ];
        list
    ]

MinimalSurfaces::noint = "Integration `1` failed."
```

Next comes the real parameterization. The function Weierstrass[f, g, z, *trans*] replaces z by *trans*, where *trans* is either $u + iv$ or $re^{i\phi}$.

```
Weierstrass[f_, g_, z_, trans_] :=
    Module[{complex},
        complex = Parameterize[f, g, z];
        ComplexExpand[ Re[complex /. z -> trans] ]
    ]
```

The two commands CartesianSurface and PolarSurface are now straightforward.

```
PolarSurface[f_, g_, z_, {r0_, r1_}, {phi0_, phi1_}, opts___] :=
    Module[{list, r, phi},
        list = Weierstrass[f, g, z, r Exp[I phi]];
        ParametricPlot3D[ Evaluate[list],
                        {r, r0, r1}, {phi, phi0, phi1}, opts ]
    ]

CartesianSurface[f_, g_, z_, {u0_, u1_}, {v0_, v1_}, opts___] :=
    Module[{list, u, v},
        list = Weierstrass[f, g, z, u + I v];
        ParametricPlot3D[ Evaluate[list],
                        {u, u0, u1}, {v, v0, v1}, opts ]
    ]
```

The use of `Evaluate` lets `ParametricPlot3D` compile the three coordinate functions before the repeated numerical evaluation.

Here is a picture of *Henneberg's surface* obtained with these functions. Plate 2 shows another two pictures of it.

```
In[1]:= PolarSurface[ 2-2/z^4, z, z,
          {0.35, 0.85}, {-Pi, Pi},
          PlotRange->All, PlotPoints->{12, 48} ];
```

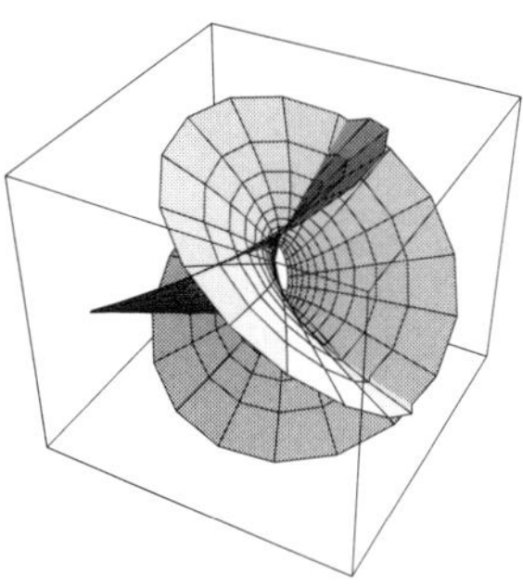

9.2.1   Graphics Extensions

Graphics functions, including `ParametricPlot3D`, can take an option `PlotPoints` to specify the number of subdivisions of the parameter ranges to use. With other iterators in *Mathematica* one can use an increment in the form $\{var, v_0, v_1, dv\}$ to give the number of iterations to perform. Sometimes, this is more convenient. It is not difficult to write additional rules for `ParametricPlot3D` to allow such increments in place of the option `PlotPoints`. The code is in the package Graphics/ParametricPlot3D.m, and is described in "Programming in *Mathematica*." We have added the same capability to `CartesianSurface` and `PolarSurface`.

Here is a picture of a minimal surface with a fivefold rotational symmetry, using increments instead of the option `PlotPoints`.

```
In[1]:= PolarSurface[ z, Sqrt[z], z,
          {0, 0.7, 0.05}, {0, 4Pi, Pi/32},
          Axes -> None, PlotRange -> All ];
```

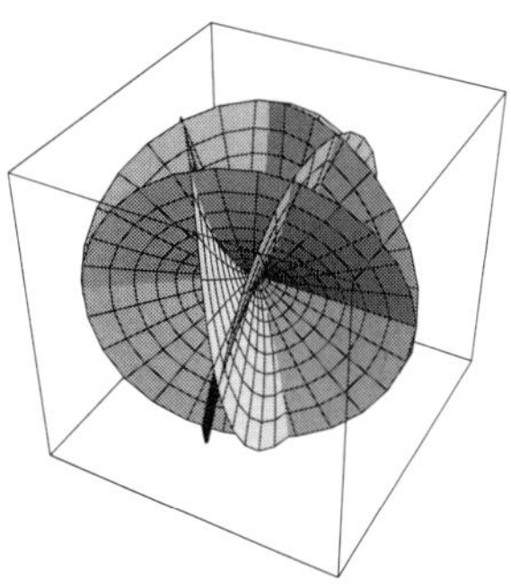

9.3 Analytic Continuations

The numerical evaluation that happens inside `ParametricPlot3D` can run into problems if the parameterization contains multivalued functions. The logarithm function is an example. *Mathematica*'s function `Log` always chooses the *principal value,* that is, the value of $\log \zeta$ with $-\pi < \mathrm{Im}(\log \zeta) \leq \pi$. If ζ varies along a path in the complex plane that crosses the negative real axis, $\log \zeta$ will have a discontinuity as ζ crosses the axis. The negative real axis is referred to as a *branch cut.*

We can see this clearly in a picture of $\mathrm{Im}(\log \zeta)$ as ζ goes once around the unit circle: $\zeta = e^{i\phi}, 0 \leq \phi \leq 2\pi$.

```
In[2]:= Plot[Im[Log[E^(I phi)]], {phi, 0, 2Pi}];
```

If the parameterization of our minimal surface involves a multivalued function with a branch cut, we will get an incorrect picture of a surface that "jumps back" onto itself. For a smooth picture, we must not choose the principal value but one of the values of the function that gives a continuous function of the parameter. Choosing a value in this way is referred to as *analytic continuation* of the multivalued function. There is no general way of computing such a value. Fortunately, we can obtain it in many cases by a simple symbolic simplification of the formula involved.

Consider again the previous example $\log e^{i\phi}$. Since the logarithm and exponential are inverses, we can simplify this to $i\phi$. Its imaginary part is simply ϕ, which is certainly a continuous function of ϕ! *Mathematica* does not perform this simplification automatically, since it would not give principal values. In our application, however, it is exactly what we want. Here are the transformation rules that help us to get an analytic continuation:

```
continuations = {
        Log[a_ b_] :> Log[a] + Log[b],
        Log[E^x_] :> x,
        Arg[a_ b_] :> Arg[a] + Arg[b],
        Arg[Exp[I phi_]] :> phi
}
```

Another kind of simplification arises from the fact that the radius in polar coordinates is always non-negative. ComplexExpand cannot deal with such knowledge. Instead, we can apply the following rules to take advantage of this fact. (Note that r can be declared positive with r/: Positive[r] = True.)

```
postrules = {
    Abs[r_?Positive] :> r,
    Arg[r_?Positive] :> 0
}
```

The continuation rules may have to be applied several times, each time calling ComplexExpand to simplify the result further. This is most easily achieved by defining a simplifier function and then computing its fixed point in a new version of Weierstrass.

```
Simplifier[e_] := ComplexExpand[e //. continuations]

Weierstrass[ f_, g_, z_, trans_ ] :=   (* in real coordinates *)
    Module[{complex, real},
        complex = Parameterize[f, g, z];
        real = Re[complex /. z -> trans];
        FixedPoint[ Simplifier, real ] //. postrules
    ]
```

We can illustrate these methods with the choice $f = \zeta$, $g = 1/\zeta$. The simple parameterization of the preceding section does not work in this case, since the result Arg[Exp[I phi] r] is not continuous.

```
In[3]:= Weierstrass[ z, 1/z, z, r Exp[I phi] ]
```

$$\text{Out[3]}= \{\frac{r^2\,\text{Cos}[2\,\text{phi}]}{2} - \frac{\text{Log}[r^2]}{2},$$

$$-\text{phi} - \text{Arg}[r] - \frac{r^2\,\text{Sin}[2\,\text{phi}]}{2},\ 2\,r\,\text{Cos}[\text{phi}]\}$$

With the new definitions, we get the correct result and can plot the surface.

```
In[1]:= Weierstrass[ z, 1/z, z, r Exp[I phi] ]
```

$$\text{Out[1]}= \{\frac{r^2\,\text{Cos}[2\,\text{phi}]}{2} - \frac{\text{Log}[r^2]}{2},\ -\text{phi} - \frac{r^2\,\text{Sin}[2\,\text{phi}]}{2},$$

$$2\,r\,\text{Cos}[\text{phi}]\}$$

Increasing the range for `phi` will add more and more curls to it. A color version of this surface is shown on Plate 2.

```
In[2]:= PolarSurface[ z, 1/z, z,
            {0.0005, 2.0005}, {-Pi, 3Pi},
            ViewPoint -> {-2.1, -1.1, 1.2}, Axes -> None,
            PlotRange -> All, PlotPoints -> {10, 60} ];
```

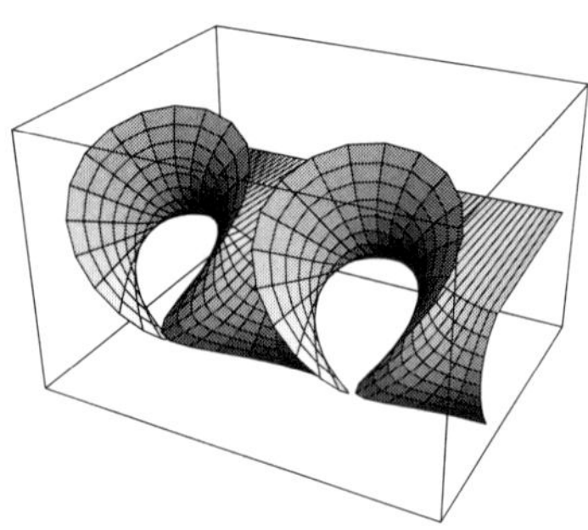

9.4 Complex Contortions

There is nothing magic about taking the real part of the complex integrals to obtain our real-valued functions. The imaginary part can do as well. Since the function f appears linearly in all three coordinates and $\mathrm{Im}(\zeta) = \mathrm{Re}(-i\zeta)$, we can obtain the imaginary part simply by replacing f by $-if$.

The result is surprising. Here are the surfaces for the real and imaginary parts with $f = e^{-\zeta}$, $g = e^{\zeta}$.

```
In[3]:= cat = CartesianSurface[
               E^-z, E^z, z, {-2, 2}, {-Pi, Pi},
               DisplayFunction->Identity ]; \
           hel = CartesianSurface[
               -I E^-z, E^z, z, {-2, 2}, {-Pi, Pi},
               DisplayFunction->Identity ];
```

The first surface is the *catenoid*; the second one is the *helicoid*.

```
In[4]:= Show[ GraphicsArray[{cat, hel}] ];
```

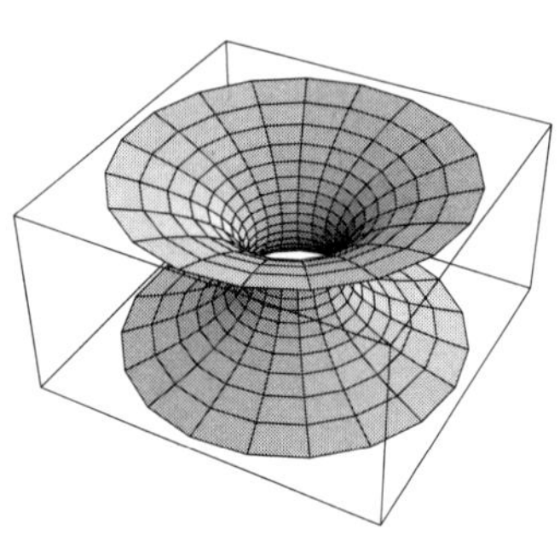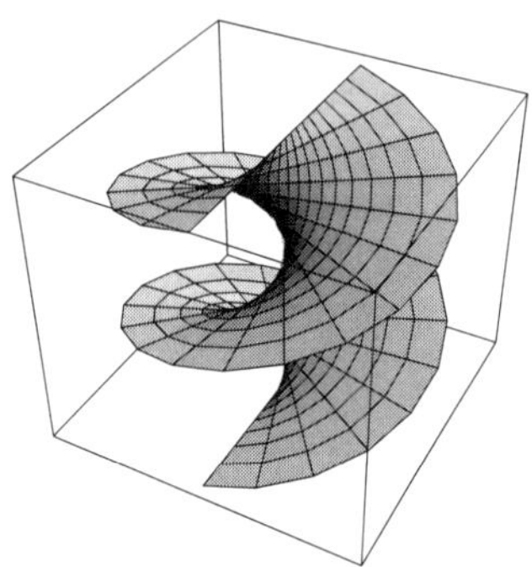

The multiplication by $-i$ corresponds to a rotation in the complex plane. Now, what about an arbitrary rotation in the complex plane? This corresponds to a multiplication of f by e^{ia} for $0 \leq a \leq 2\pi$. And while we're at it, why not animate the results? Animation doesn't quite work in a printed medium, so we'll have to do it with a static collection of frames, using FlipBookAnimation.m, another package from "Programming in *Mathematica*." The chapter title shows the first quadrant, $0 \leq a \leq \pi/2$, showing the *smooth* transition of the catenoid into the helicoid! It was produced with this command:

```
Animate[
    CartesianSurface[Exp[I arg] E^-z, E^z, z,
        {-2, 2, 4/20}, {-Pi, Pi, Pi/14},
        PlotRange->{{-7.5, 7.5}, {-7.5, 7.5}, {-7.5, 7.5}},
        Boxed->False, Axes->None],
    {arg, 0, Pi/2}, Frames->9, Closed->False
];
```

The animation command for the chapter title graphics

The Notebook MinimalExamples.ma contains the code for producing the true animation.

9.5 Bonus Pictures

Some more surfaces follow. I do not know whether any of these, or the curly surface on page 156, is already known and has a name. I welcome comments from readers who have seen any of them before.

The first example has a threefold symmetry.

```
In[1]:= PolarSurface[ 1, Sqrt[z], z,
                {0.01, 1.01, 1/8}, {0, 4Pi, Pi/12}
        ];
```

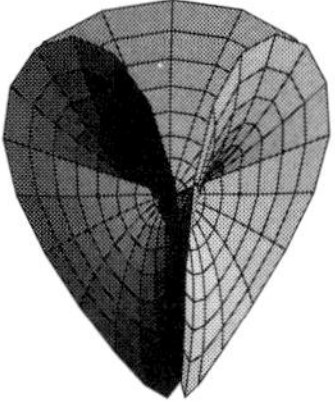

The second example is singular at the origin.

```
In[2]:= CartesianSurface[
            Tan[z], Tan[z], z, {-1.1, 1.1, 2.2/30},
            {-1.6, 1.6, 3.2/50},
            ViewPoint->{2.2, -1.6, 1.8}, PlotRange->All ];
```

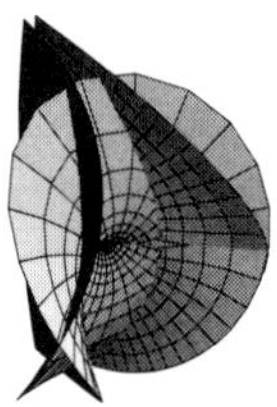

If you want to hunt for new surfaces, keep in mind that not every pair of f and g gives a different surface. For example, here is a different parameterization of the catenoid.

```
In[3]:= PolarSurface[1, 1/z, z, {1/5, 5}, {-Pi, Pi}];
```

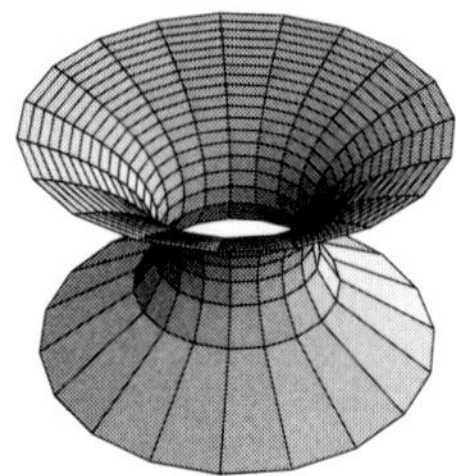

9.6 Further Reading

Dickson's article [11] contains further references to the mathematical theory of minimal surfaces. The complete complex analysis for the Enneper-Weierstrass parameterization can be found in [38]. A recommended text on differential geometry is [12]. A classical introduction to complex analysis is [18].

The floppy disk contains the package MinimalSurfaces.m and a Notebook MinimalExamples.ma with all the examples presented here (and some more). It contains only the commands for rendering the surfaces and animations, not their output, due to disk space limitations.

9.7   The Complete Code of MinimalSurfaces.m

The two functions `CartesianParameters` and `PolarParameters` were not described in this article. They return the real parameterization as a pure function instead of producing the picture directly.

```
BeginPackage["MinimalSurfaces`", "Utilities`FilterOptions`"]

PolarSurface::usage = "PolarSurface[f, g, z, {r0:0, r1, (dr)}, {phi0, phi1, (dphi)},
    options...] plots a minimal surface with Enneper-Weierstrass functions f
    and g in polar coordinates. The phi range defaults to {-Pi, Pi}."

CartesianSurface::usage = "CartesianSurface[f, g, z, {u0, u1, (du)}, {v0, v1, (dv)},
    options...] plots a minimal surface with Enneper-Weierstrass functions f
    and g in cartesian coordinates."

Parameterize::usage = "Parameterize[f, g, z] gives
    Integrate[{f - f g^2, I(f + f g^2), 2 f g}, z]
    (Weierstrass-parameterization)."

PolarParameters::usage = "PolarParameters[f, g, z] gives the polar parameterization
    of Integrate[{f - f g^2, I(f + f g^2), 2 f g}, z] as a function of r and phi."

CartesianParameters::usage = "CartesianParameters[f, g, z] gives the cartesian
    parameterization of Integrate[{f - f g^2, I(f + f g^2), 2 f g}, z]
    as a function of u and v ."

Weierstrass::usage = "Weierstrass[f, g, z, trans] gives parameters
    according to the transformation trans."

MinimalSurfaces::noint = "Integration `1` failed."

Begin["`Private`"]

Parameterize[ f_, g_, z_ ] :=                    (* the integration *)
   Module[{list},
       list = Integrate[{f - f g^2, I(f + f g^2), 2 f g}, z];
       If[ !FreeQ[list, Integrate],
               Message[MinimalSurfaces::noint, list]; Abort[] ];
       list
   ]

Weierstrass[ f_, g_, z_, trans_ ] :=     (* in real coordinates *)
   Module[{complex, real},
       complex = Parameterize[f, g, z];
       real = Re[complex /. z -> trans];
       FixedPoint[ Simplifier, real ] //. postrules
   ]

PolarSurface[ f_, g_, z_, {r0_:0, r1_, dr_:Automatic},
            {p0_, p1_, dp_:Automatic}, opts___ ] :=
   Module[{list, r, p},
       r /: Positive[r] = True;    (* declare local r positive *)
       list = Weierstrass[f, g, z, r Exp[I p]];
```

```mathematica
        DoSurface[ list, {r, r0, r1, dr}, {p, p0, p1, dp}, opts ]
    ]

(* default phi range *)
PolarSurface[ f_, g_, z_, rrange_List, opts___ ] :=
        PolarSurface[f, g, z, rrange, {-Pi, Pi}, opts]

CartesianSurface[ f_, g_, z_, {u0_, u1_, du_:Automatic},
                  {v0_, v1_, dv_:Automatic}, opts___ ] :=
    Module[{list, u, v},
        list = Weierstrass[f, g, z, u + I v];
        DoSurface[ list, {u, u0, u1, du}, {v, v0, v1, dv}, opts ]
    ]

(* the graphics details: PlotPoints *)
DoSurface[ coord_, {u_, u0_, u1_, du_}, {v_, v0_, v1_, dv_}, opts___ ] :=
    Module[{ndu = N[du], ndv = N[dv], plotpoints, pu, pv},
        plotpoints = PlotPoints /. {opts} /. Options[ParametricPlot3D];
        If[Head[plotpoints] =!= List,     (* a single value *)
                {pu, pv} = {plotpoints, plotpoints},
                {pu, pv} = plotpoints      (* two values *)
        ];
        If[ NumberQ[ndu], pu = Round[N[1 + (u1 - u0)/ndu]] ];
        If[ NumberQ[ndv], pv = Round[N[1 + (v1 - v0)/ndv]] ];
        ParametricPlot3D[
            Evaluate[N[coord]], {u, u0, u1}, {v, v0, v1},
            PlotPoints -> {pu, pv},
            Evaluate[FilterOptions[ParametricPlot3D, opts]]]
    ]

(* rules for analytic continuations of log, exp and arg *)
continuations = {
        Log[a_ b_] :> Log[a] + Log[b],
        Log[E^x_] :> x,
        Arg[a_ b_] :> Arg[a] + Arg[b],
        Arg[Exp[I phi_]] :> phi
}

(* further simplifications *)
postrules = {
        Abs[r_?Positive] :> r,
        Arg[r_?Positive] :> 0
}

(* the canonical simplifier for the real part *)

Simplifier[e_] := ComplexExpand[e //. continuations]

(* "Function @@ {e}" generates a pure function with body e *)

PolarParameters[f_, g_, z_] :=
    Module[{r, phi, w},
        r /: Positive[r] = True;     (* declare local r positive *)
        w = Weierstrass[f, g, z, r Exp[I phi]];
        Function @@ {w /. {r -> #1, phi -> #2}}
```

```
    ]

CartesianParameters[f_, g_, z_] :=
    Module[{u, v, w},
        w = Weierstrass[f, g, z, u + I v];
        Function @@ {w /. {u -> #1, v -> #2}}
    ]

End[ ]

Protect[PolarSurface, CartesianSurface, Parameterize,
        PolarParameters, CartesianParameters, Weierstrass]

EndPackage[ ]
```

Chapter Ten
System Programming

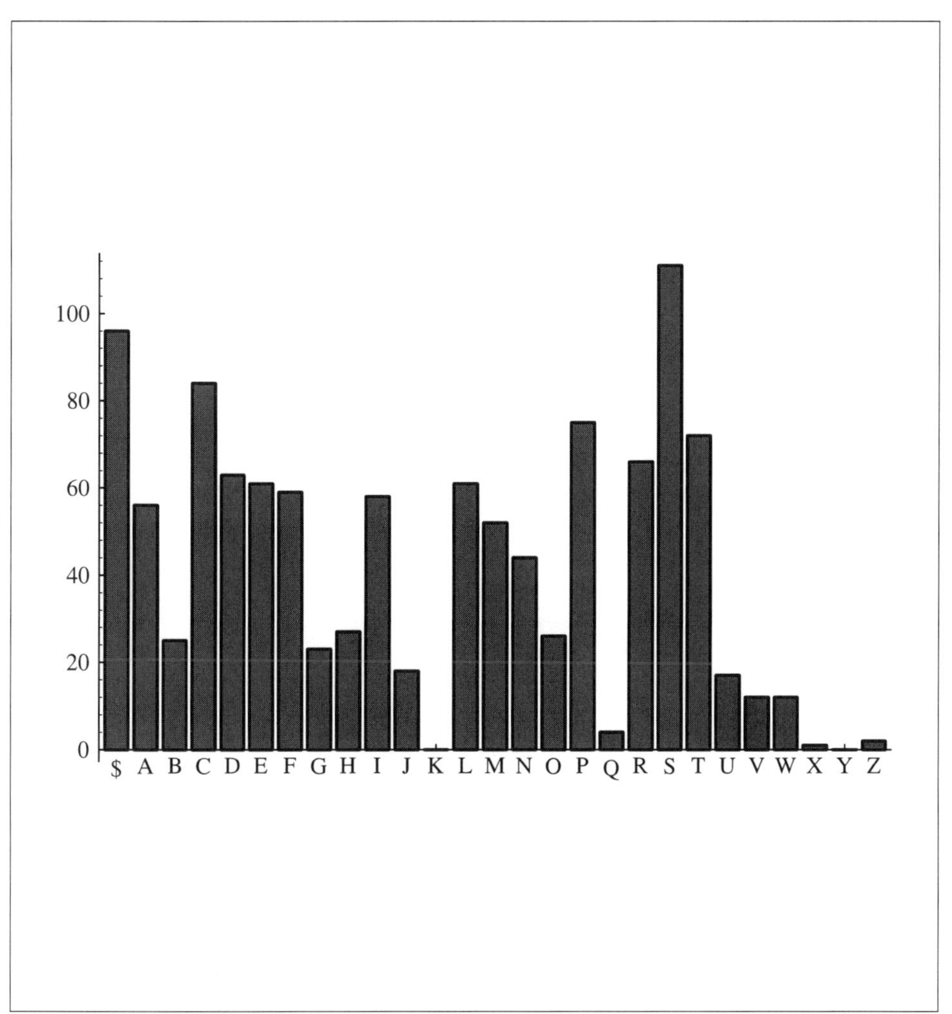

System programming refers to procedures and tools for maintaining the computer system itself, as opposed to application programming. *Mathematica* is mainly used for programming applications, i.e., solving mathematical problems. In this article we develop tools for looking at *Mathematica* itself. Such tools are useful in program development. For example, we can find all symbols in a context having no documentation, revealing possible omissions or possibly bad package design.

As usual when a system is used to look upon itself, problems of self-referentiality arise. One example is the difficulty of writing a manual for a text-formatting language in that very language. Printing special characters (such as the backslash \ in TeX) requires careful quotation or escape mechanisms to prevent them from acting in the usual way. In *Mathematica* the main problem is unwanted evaluation of expressions we want to treat as data.

About the illustration overleaf:
The frequencies of initial letters of all built-in symbols in *Mathematica*. The commands for this pictures are in Pictures.m. The barcharts are from the standard package Graphics‘Graphics‘.

10.1 Working with Symbols

Suppose we want a list of all symbols present in a given *Mathematica* context. Interactively, we can obtain this information by typing *?Context`**.  The symbol names are printed on the terminal (or in the Notebook), but not in a form that we could use further.  The construction `Names["Context`*"]` is closer to what we want (note that the argument must be a string), but the list it returns is made of strings, not of the symbols proper.

Working with the symbols themselves is tricky because almost every *Mathematica* expression tries to evaluate them. Thus, if `Names` returned a list of symbols, the end result would be a list of the symbol values—not what we want. Before going further, then, we need some tools for working with lists of symbols.

One possibility is to wrap `Hold` or `HoldForm` around all symbols, to protect them from evaluation, but the wrapper would get in the way later. Instead we will use a different *container* in place of a list. A symbol with attribute `HoldAll` can be used as head of an expression containing our symbols as elements. In this way they are again protected from evaluation. We call this container `HoldList`. The implementation is trivial:

```
SetAttributes[HoldList, HoldAll]
```

We can easily have these held lists print like ordinary lists:

```
Format[holdlist_HoldList] := List @@ HoldForm /@ holdlist
```

Here we do use the wrapper `HoldForm`, but only for formatting. After it has been applied to the elements of `holdlist` we can now safely replace the head `HoldList` by the ordinary `List`.

To see how all of this works, we create some global symbols, some of them having values. A good way to catch any attempt at evaluating a symbol is to give it a delayed value that causes a warning to be printed when it is evaluated.

These three symbols will serve as test cases for the following code.

```
In[1]:= a; v1 = 55; guard := Print["gotcha!"]
```

We now try to construct a held list of all global symbols. We start by using the `Names[ ]` command, mentioned earlier. Although the standard output form is ambiguous, the elements of the resulting list are really strings.

This returns a list of all matching symbols as strings. The default output form of strings omits the enclosing quotation marks, so this is not obvious.

```
In[2]:= Names["Global`*"]

Out[2]= {a, guard, v1}
```

A look at the input form makes this clear.

```
In[3]:= InputForm[%]

Out[3]//InputForm= {"a", "guard", "v1"}
```

How can we transform this into a held list of the symbols themselves? First, we have to turn the strings into symbols. `ToExpression` does this, but since the symbols are part of a list, they would be evaluated afterward. However, there is a function `ToHeldExpression` that wraps its result in `Hold`.

`ToHeldExpression` wraps its result in `Hold`.	`In[4]:= ToHeldExpression /@ %`
	
	`Out[4]= {Hold[a], Hold[guard], Hold[v1]}`

Now we can replace the list by a held list. Observe that the special output form defined earlier makes the result print as a regular list.	`In[5]:= HoldList @@ %`
	
	`Out[5]= {Hold[a], Hold[guard], Hold[v1]}`

And finally we get rid of the now unnecessary `Hold` wrappers. This is done through a replacement rule, which does not evaluate its result but works inside held expressions.	`In[6]:= % /. Hold[sym_] :> sym`
	
	`Out[6]= {a, guard, v1}`

Note that we never saw the message "gotcha!" during these operations. Nevertheless we now have a list of the symbols themselves, as the input form shows.	`In[7]:= InputForm[%]`
	
	`Out[7]//InputForm= HoldList[a, guard, v1]`

We collect these steps in a function `SymbolsInContext["Context`"]`.

```
SymbolsInContext[context_String] :=
    HoldList @@ ToHeldExpression /@ Names[context <> "**"] /. Hold[sym_] :> sym
```

10.2 Housekeeping

`Select` works on held lists in the same way it does on ordinary lists. You can use it, for instance, to get a list of all global symbols that have a value:

```
In[8]:= Select[ SymbolsInContext["Global`"], ValueQ ]
Out[8]= {guard, v1}
```

The definition of `ValueQ` in version 2 of *Mathematica* has the undesirable side effect of evaluating its argument in order to find out whether it has a value. For symbols there is an easier way to find out (provided they don't have the attribute `ReadProtect`). A better definition of `ValueQ`, used here, is part of the package SystemProgramming.m.

```
protected = Unprotect[ ValueQ ]

ValueQ[sym_Symbol] /; !HasAttribute[sym, ReadProtected] := Length[OwnValues[sym]] > 0

Protect[ Evaluate[protected] ]
```

Checking a symbol without evaluating it

Here is what happens with the built-in definition of `ValueQ`.

```
In[1]:= ValueQ[guard]

gotcha!

Out[1]= True
```

Using the same technique, we can apply other useful predicates to a list of symbols. Table 10.2–1 lists some of these predicates.

`HasOptions[`*symbol*`]`	True if *symbol* has options.
`UndocumentedQ[`*symbol*`]`	True if *symbol* does not have a usage message.
`UnprotectedQ[`*symbol*`]`	True if *symbol* is unprotected.
`ValueQ[`*symbol*`]`	True if *symbol* has a value.

Table 10.2–1: Some useful predicates

The implementation of these predicates has a few subtleties, which is expected in system programming.

```
SetAttributes[ {HasOptions, UndocumentedQ, UnprotectedQ, HasAttribute, HasOption},
          HoldFirst ]

HasOptions[sym_Symbol]  := Length[Options[Unevaluated[sym]]] > 0
UndocumentedQ[sym_Symbol] := Head[sym::usage] =!= String
HasAttribute[sym_Symbol, attr_] := MemberQ[Attributes[sym], attr]
UnprotectedQ[sym_Symbol] := !HasAttribute[sym, Protected]
HasOption[sym_Symbol, opt_] := MemberQ[First /@ Options[Unevaluated[sym]], opt]
```

The definition of the symbol predicates

First, all the predicates need the attribute `HoldFirst` to prevent evaluation of their first argument (the symbol to which they are applied). To decide if a symbol has options, we use `Options`, but this function itself does not have the attribute `HoldFirst`; therefore we wrap its argument in `Unevaluated`.

Similarly, a symbol is undocumented if it has no usage message. The construct *symbol*`::usage` (internally `MessageName[`*symbol*`, "usage"]`) evaluates to the message (a string), if the message exists. We can test this by looking at its head. `MessageName` already has the attribute `HoldFirst`.

Finally, to see whether a symbol is unprotected, we check if `Protected` is in its list of attributes. The auxiliary function `HasAttribute[`*symbol*`, `*attribute*`]` returns `True` if *symbol* has the given attribute. This is the case if the attribute is a member of the list of attributes. `HasOption[`*symbol*`, `*option*`]` is entirely similar.

We can now apply these commands to the system symbols. If you are developing your own package, you might want to do the same with its symbols. For the following examples we have set `$PrePrint = Short` to avoid excessive amounts of output.

Here is the list of all system symbols.

```
In[1]:= system = SymbolsInContext["System`"]
Out[1]= {Abort, $Aborted, AbortProtect, Above, Abs,
    AbsoluteDashing, A<<14>>ze, <<1115>>, Xor, ZeroTest, Zeta}
```

This is the list of system symbols having options. There are quite a number of them.

```
In[2]:= Select[system, HasOptions]
Out[2]= {AccountingForm, AlgebraicRules, Apart,
    ApartSquareFree, Apply, <<129>>, Union, Variables, Zeta}
```

The complement of this list is the list of all symbols that do not have options.

```
In[3]:= Complement[system, %]
Out[3]= {Abort, $Aborted, AbortProtect, Above, Abs,
    AbsoluteDashing, <<979>>, WynnDegree, Xor, ZeroTest}
```

Here is the (short) list of undocumented system symbols. It is not a good idea to use any of them.

```
In[4]:= Select[system, UndocumentedQ]
Out[4]= {$AlwaysUsePrivateColorMap, $AnimationFunction,
    CallPacket, $Cli<<11>>ow, <<28>>, UnAlias, UseDump}
```

And, finally, the list of unprotected symbols. These are the ones for which Wolfram Research thinks it's relatively safe or common to define values (but some symbols are on this list by oversight, so beware!).

```
In[5]:= Select[system, UnprotectedQ]
Out[5]= {$AlwaysUsePrivateColorMap, $AnimationFunction,
    $AutoLoad, $BatchInput, <<78>>, Uninstall, $Urgent}
```

On systems where usage messages are not preloaded (the so-called *thin kernel*), the command `Select[system, UndocumentedQ]` will take quite a while since all messages are read in at this point.

10.2.1 Convenience

It is not very convenient to have to define a different test function for each property a symbol might have. A general construct using a pure function as an argument would be much simpler. For example, we could try

```
Select[system, HasOption[#, PlotPoints]&]
```

to find all commands that accept the option `PlotPoints`. But there is a problem with this: pure functions do not have attributes, so each symbol would be evaluated before being passed to `HasOption`.

Now, a pure function with named arguments can have attributes, so that

```
Select[system, Function[sym, HasOption[sym, PlotPoints], {HoldFirst}]]
```

has the desired effect. However, this notation is pretty obscure; it would be better to stick to pure functions with unnamed arguments, using the # and & notation.

Our solution is to redefine the action of `Select` on a held list. If the first argument of `Select` is a held list and its second argument is a pure function with unnamed arguments, we automatically make the pure function have the attribute `HoldFirst`:

```
HoldList/: Select[ holdlist_HoldList, body_& ] :=
    With[{newf = Function[sym, bd, {HoldFirst}] /. bd :> body /. # :> sym},
        Select[ holdlist, newf ]
    ]
```

The subtleties explained: The definition belongs to `HoldList`, not to the (protected) system function `Select`. The pattern `body_&` matches a pure function with unnamed arguments, that is, something of the form `Function[`*body*`]`, but not `Function[`*args, body*`]`. We turn this into a pure function with a named argument by replacing all occurrences of # by `sym`. To prevent renaming of variables (as lexical scoping would demand), we use another replacement to "smuggle" the body of the pure function inside. Finally, the built-in code of `Select` takes over, since our rule no longer matches.

Now we can write predicates on the fly and test any desired property easily. Here are all locked symbols.

```
In[6]:= Select[system, HasAttribute[#, Locked]&]
Out[6]= {$Aborted, $BatchOutput, $CommandLine,
    $CreationDate, $DumpDates, <<22>>, $Version,
    $VersionNumber}
```

And here are the symbols with more than 10 options (this includes most graphics commands).

```
In[7]:= Select[ system,
                Length[Options[Unevaluated[#]]] > 10& ]
Out[7]= {ContourGraphics, ContourPlot, DensityGraphics,
    DensityPlot, <<7>>s, <<9>>, Plot3D, SurfaceGraphics}
```

10.3 Applying Functions to Symbols

Besides selecting symbols based on a property we can also apply functions to these symbols. A function *f* is applied to the elements of a list with `Map[`*f, list*`]`. This works also for our held lists, but the result of the application is not evaluated, so the functions are never called. Having applied a function to the symbols, we no longer want the protection offered by `HoldList`. The easiest way to force evaluation is to replace the container `HoldList` by `List` after the mapping.

Here again is our held list of some sample symbols:	`In[1]:= HoldList[a, v1, guard]`
	
	`Out[1]= {a, v1, guard}`

Mapping a function over this list does nothing at first.	`In[2]:= Map[ ValueQ, % ]`
	
	`Out[2]= {ValueQ[a], ValueQ[v1], ValueQ[guard]}`

Now we let the evaluator loose.	`In[3]:= List @@ %`
	
	`Out[3]= {False, True, True}`

Again, this is a programming paradigm that we should make automatic. A simple rule for mapping of held lists does it. We also do this for the function `Through` used later on. Also, we give simple pure functions the attribute `HoldAll`, as above.

```
HoldList/: Map[body_&, holdlist_HoldList, args___] :=
    With[{newf = Function[sym, bd, {HoldFirst}] /. bd :> body /. # :> sym},
        Map[ newf, holdlist, args ]
    ]

HoldList/: Map[f_, HoldList[elements___], args___] :=
        List @@ Map[f, Hold[elements], args]

HoldList/: Through[HoldList[elements___][args___]] :=
        List @@ Through[Hold[elements][args]]
```

Functional operations on held lists

The only subtlety here is the replacement of the head `HoldList` by `Hold` before the built-in code for `Map` is called (any head with attribute `HoldAll` would do). I leave it to the reader to find out why this is done.

For an advanced application, we generate a list of the names of all options appearing in *Mathematica*.

First, we find the options of all symbols.	`In[1]:= Options[Unevaluated[#]]& /@ system`
	
	`Out[1]= {{}, {}, {}, {}, {}, {}, {}, {}, {}, {<<9>>},`
	
	`    <<1110>>, {}, {}, {}, {}, {IncludeSingularTerm -> False}}`

Now we replace the options given in the form *name -> value* by their names alone.	`In[2]:= Map[ First, %, {2} ]`
	
	`Out[2]= {{}, {}, {}, {}, {}, {}, {}, {}, {}, <<1111>>, {},`
	
	`    {}, {}, {}, {IncludeSingularTerm}}`

Finally, we take the union of all option lists, merging duplicates and sorting them.	`In[3]:= Union @@ %`
	
	`Out[3]= {AccuracyGoal, AmbientLight, Analytic,`
	
	`    AnchoredSearch, AspectRatio, <<155>>, WynnDegree,`
	
	`    ZeroTest}`

10.4   A Crash Test

The little known function

$$\texttt{Through[} \{h_1, h_2, \ldots, h_n\}\texttt{[}args\ldots\texttt{]]}$$

takes the function call inside the list to produce

$$\{h_1\texttt{[}args\ldots\texttt{], } h_2\texttt{[}args\ldots\texttt{], }\ldots\texttt{, } h_n\texttt{[}args\ldots\texttt{]}\}$$

effectively calling all functions $h_1, h_2, \ldots, h_n$ with the same arguments *args*. One effective test of
system robustness is calling all built-in functions with "crazy" arguments. Even calling them with
no arguments produced quite a few crashes in version 1.2 of *Mathematica*. With tests like these,
they were systematically found and fixed. When you do this, you will get tons of error messages,
mostly about the wrong number of arguments. Some functions should be excluded from such tests,
among them all input functions (like `Input` or `Edit`), debugging functions (`Debug`, `Dialog` and
friends), and also `Exit` and `Quit`, for obvious reasons. A list of such dangerous symbols is in
the package SystemProgramming.m under the name `dangerousSymbols`:

```
dangerousSymbols = HoldList[
      Abort, Exit, Quit, Debug, Dialog, Interrupt,
      Edit, EditIn, EditDef, EditDefinition,
      Trace, TraceDialog, TracePrint, TraceScan, $TracePostAction, $TracePreAction,
      Read, Input, InputString, Interrupt, On, Off,
      Throw, $Throw,
      In, Out
   ]
```

Here is this test applied to the package SystemProgramming.m itself.

```
In[1]:= <<SystemProgramming.m
```

First we generate a list of all exported symbols.

```
In[2]:= testing = SymbolsInContext["SystemProgramming`"]

Out[2]= {dangerousSymbols, globalSymbols, HasAttribute,

   HasOption, HasOptions, HoldList, SetAllOptions,

   SymbolsInContext, systemSymbols, UndocumentedQ,

   UnprotectedQ}
```

Now we select all symbols that do
not have values. Only they can be
used as functions.

```
In[3]:= funcs = Select[ %, !ValueQ[#]& ]

Out[3]= {HasAttribute, HasOption, HasOptions, HoldList,

   SetAllOptions, SymbolsInContext, UndocumentedQ,

   UnprotectedQ}
```

Next, we call all of them without arguments. See the listing of SystemProgramming.m for how the error message was generated.

```
In[4]:= Through[funcs[]]

SymbolsInContext::argx:
   SymbolsInContext called with 0
      arguments; 1 argument is expected.

Out[4]= {HasAttribute[], HasOption[], HasOptions[], {},

    Null, SymbolsInContext[], UndocumentedQ[], UnprotectedQ[]}
```

When debugging your own packages, you should watch for results where any error in the arguments causes other, seemingly unrelated, errors later on. In these cases you should add your own argument checking. How this is done is explained in [28].

10.5   Options

Many options are used in more than one function, especially with graphics. Here is a way of setting the default value for all functions that understand the option. We use the tools just developed to find all symbols that know about the option and then reset it to a new value.

All of these functions know the option **Axes**.

```
In[1]:= Select[system, HasOption[#, Axes]&]

Out[1]= {ContourGraphics, ContourPlot, DensityGraphics,

    DensityPlot, Graphics, GraphicsArray, Graphics3D,

    ListContourPlot, ListDensityPlot, ListPlot, ListPlot3D,

    ParametricPlot, ParametricPlot3D, Plot, Plot3D,

    SurfaceGraphics}
```

These are the default values of the option **Axes** in all cases.

```
In[2]:= Axes /. Options /@ %

Out[2]= {False, False, False, False, False, False, False,

    False, False, Automatic, True, Automatic, None,

    Automatic, True, False}
```

Now we set all of these default values to **False** with our function **SetAllOptions**.

```
In[3]:= SetAllOptions[Axes->False]
```

Indeed:

```
In[4]:= Axes /. Options /@ %%%

Out[4]= {False, False, False, False, False, False, False,

    False, False, False, False, False, False, False,

    False}
```

Here is the code. Note how we make sure that only valid rules (of two sorts) are accepted as arguments and how the function is made to work with any number of arguments.

```
SetAllOptions[arg: (opt_ -> _) | (opt_ :> _)] :=
    Module[{syms},
        syms = HoldList @@ ToHeldExpression /@ Names["*"] /. Hold[sym_] :> sym;
    syms = Select[syms, HasOption[#, opt]&];
    Map[ SetOptions[#, arg]&, syms ];
    ]

SetAllOptions[arg:(_Rule|_RuleDelayed)...] := (SetAllOptions /@ {arg}; )
```

Setting options for several functions at once

10.6 Conclusions

Mathematica's programming language is well suited for introspection: looking back onto itself. The usual problems with self-referentiality appear mainly as unwanted evaluations of expressions that we want to treat as data. With suitable data types these problems can be solved.

To test your understanding of these ideas, you may want to solve some of these exercises:

- Which function has the largest number of options?
  Which option is used in the largest number of functions?

- Make a histogram showing how many symbols start with each letter of the alphabet.

- Which symbol has the longest name?

- Are there any system symbols without any attributes or messages?

The floppy disk contains the package SystemProgramming.m with the definition of all the functions described.

10.7 The Complete Code of SystemProgramming.m

```
BeginPackage["SystemProgramming`"]

SymbolsInContext::usage = "SymbolsInContext[\"context`\"] gives a
    HoldList of all symbols in context."

HoldList::usage = "HoldList[elements..] is a list that does
    not evaluate its elements. Mapping a function over its
    elements turns it into an ordinary list."

systemSymbols::usage = "systemSymbols is a HoldList of all system symbols."
globalSymbols::usage = "globalSymbols is a HoldList of all global
    (user-defined) symbols."

dangerousSymbols::usage = "dangerousSymbols is a HoldList of system symbols
    that should probably not be called as functions."

HasOptions::usage = "HasOptions[symbol] is True if symbol has any options."
UndocumentedQ::usage = "UndocumentedQ[symbol] is True if symbol
    has no usage message."
UnprotectedQ::usage = "UnprotectedQ[symbol] is True if symbol is unprotected."
HasAttribute::usage = "HasAttribute[symbol, attr] is True if
    symbol has the attribute attr."
HasOption::usage = "HasOption[symbol, opt] is True if symbol has the given option opt."

SetAllOptions::usage = "SetAllOptions[name1->value1, name2->value2, ...] sets
    the specified default values for all symbols that know about the options."

Begin["`Private`"]

SetAttributes[HoldList, HoldAll]

SymbolsInContext[context_String] :=
    HoldList @@ ToHeldExpression /@ Names[context <> "**"] /. Hold[sym_] :> sym

(* argument errors *)

SymbolsInContext[arg_] := nothing /;
    Message[SymbolsInContext::string, 1, SymbolsInContext]
SymbolsInContext[args___] := nothing /; Length[{args}] != 1 &&
    Message[SymbolsInContext::argx, SymbolsInContext, Length[{args}]]

Format[holdlist_HoldList] := List @@ HoldForm /@ holdlist

(* make simple pure functions HoldAll in Select and Map
   with HoldList *)

HoldList/: Select[ holdlist_HoldList, body_& ] :=
    With[{newf = Function[sym, bd, {HoldFirst}] /. bd :> body /. # :> sym},
        Select[ holdlist, newf ]
    ]

HoldList/: Map[body_&, holdlist_HoldList, args___] :=
    With[{newf = Function[sym, bd, {HoldFirst}] /. bd :> body /. # :> sym},
        Map[ newf, holdlist, args ]
    ]
```

```
(* Map and Through turn HoldList into ordinary List *)

HoldList/: Map[f_, HoldList[elements___], args___] :=
           List @@ Map[f, Hold[elements], args]

HoldList/: Through[HoldList[elements___][args___]] :=
           List @@ Through[Hold[elements][args]]

systemSymbols := SymbolsInContext["System`"]
globalSymbols := SymbolsInContext["Global`"]

(* symbols that should not be called as functions in a test *)

dangerousSymbols = HoldList[
    Abort, Exit, Quit, Debug, Dialog, Interrupt,
    Edit, EditIn, EditDef, EditDefinition,
    Trace, TraceDialog, TracePrint, TraceScan, $TracePostAction, $TracePreAction,
    Read, Input, InputString, Interrupt, On, Off,
    Throw, $Throw, In, Out
    ]

(* useful predicates *)

SetAttributes[ {HasOptions, UndocumentedQ, UnprotectedQ,
               HasAttribute, HasOption}, HoldFirst ]

HasOptions[sym_Symbol] := Length[Options[Unevaluated[sym]]] > 0
UndocumentedQ[sym_Symbol] := Head[sym::usage] =!= String
HasAttribute[sym_Symbol, attr_] := MemberQ[Attributes[sym], attr]
HasOption[sym_Symbol, opt_] := MemberQ[First /@ Options[Unevaluated[sym]], opt]
UnprotectedQ[sym_Symbol] := !HasAttribute[sym, Protected]

(* set option defaults for several symbols at once *)

SetAllOptions[arg: (opt_ -> _) | (opt_ :> _)] :=
    Module[{syms},
        syms = HoldList @@ ToHeldExpression /@ Names["*"] /. Hold[sym_] :> sym;
    syms = Select[syms, HasOption[#, opt]&];
    Map[ SetOptions[#, arg]&, syms ];
    ]

SetAllOptions[arg:(_Rule|_RuleDelayed)...] := (SetAllOptions /@ {arg}; )

(* improve ValueQ for symbols *)

protected = Unprotect[ ValueQ ]

ValueQ[sym_Symbol] /; !HasAttribute[sym, ReadProtected] := Length[OwnValues[sym]] > 0

Protect[ Evaluate[protected] ]

End[ ]

Protect[ SymbolsInContext, HoldList, SetAllOptions, HasOptions,
        UndocumentedQ, HasAttribute, HasOption, UnprotectedQ ]

EndPackage[ ]
```

Appendices

Bibliography

[1]   H. Abelson and G. J. Sussman. *Structure and Interpretation of Computer Programs*. The MIT Press, Cambridge, Mass., 1985.

[2]   *AVS User's Guide, Release 4*, May 1992.

[3]   G. M. Birtwistle, O.-J. Dahl, B. Myhrhaug, and K. Nygaard. *SIMULA Begin*. Studentlitteratur Sweden, 1979.

[4]   Grady Booch. *Object Oriented Design with Applications*. The Benjamin/Cummings Publishing Company, 1991.

[5]   Walter S. Brainerd and Lawrence H. Landweber. *Theory of Computation*. John Wiley&Sons, 1974.

[6]   Timothy Budd. *An Introduction to Object-Oriented Programming*. Addison-Wesley, 1991.

[7]   Georg Cantor. uber eine Eigenschaft des Inbegriffs aller reellen algebraischen Zahlen. *Crelles Journal f. Mathematik*, 77:258–262, 1874.

[8]   Brad J. Cox and Andrew J. Novobilski. *Object-Oriented Programming*. Addison-Wesley, second edition, 1991.

[9]   H. S. M. Coxeter, Patrick du Val, H. T. Flather, and J. F. Petrie. *The Fifty-Nine Icosahedra*. Springer-Verlag, 1982. Originally published by Univ. of Toronto Press, 1938.

[10] Nachum Dershovitz and Jean-Pierre Jouannaud. Rewrite systems. In van Leeuwen [46].

[11] Stewart Dickson. Minimal surfaces. *The Mathematica Journal*, 1(1), 1990.

[12] Manfredo Perdiago do Carmo. *Differential Geometry of Curves and Surfaces*. Prentice-Hall, 1976.

[13] M. Gardner. Mathematical games. *Scientific American*, December 1976.

[14] J. A. Goguen, J. W. Thatcher, and E. G. Wagner. An initial algebra approach to the specification, correctness, and implementation of abstract data types. In Raymond T. Yeh, editor, *Current Trends in Programming Methodology IV*. Prentice Hall, 1978.

[15] J. A. Goguen, J. W. Thatcher, E. G. Wagner, and J. B. Wright. Abstract data types and correctness of data representations. In Allen Klinger et al., editors, *Proceedings of the Conference on Computer Graphics, Pattern Recognition and Data Structure*, 1975.

[16] Adele Goldberg. *Smalltalk-80*. Addison-Wesley, second edition, 1989.

[17] Peter Henrici. *Elements of Numerical Analysis*. John Wiley&Sons, 1964.

[18] Peter Henrici. *Applied and Computational Complex Analysis*, volume 2 of *Wiley Classics Library*. John Wiley&Sons, 1969.

[19] Rolf Herken, editor. *The Universal Turing Machine: A Half Century Survey*. Kammerer & Unverzagt, Hamburg–Berlin, 1988.

[20] Douglas R. Hofstadter. *Godel, Escher, Bach: An Eternal Golden Braid*. Penguin Books, 1980.

[21] D. E. Knuth. *Fundamental Algorithms*, volume 1 of *The Art of Computer Programming*. Addison-Wesley, second edition, 1973.

[22] D. E. Knuth. *Seminumerical Algorithms*, volume 2 of *The Art of Computer Programming*. Addison-Wesley, second edition, 1981.

[23] Donald E. Knuth and P. B. Bendix. Simple word problems in universal algebras. In J. Leech, editor, *Computational Problems in Abstract Algebra*. Pergamon Press, 1970.

[24] Donald E. Knuth, Ronald L. Graham, and Oren Patashnik. *Concrete Mathematics*. Addison-Wesley, 1989.

[25] Craig E. Kolb. *Rayshade User's Guide and Reference Manual*, January 10, 1992.

[26] Leslie Lamport. *LaTeX: A Document Preparation System*. Addison-Wesley, 1986.

[27] Leslie Lamport. *MakeIndex: An Index Processor for LaTeX*, February 17, 1987.

[28] Roman E. Maeder. *Programming in Mathematica*. Addison-Wesley, second edition, 1991.

[29] Roman E. Maeder. *Mathematica* as a programming language. *Dr. Dobbs Journal*, February 1992.

[30] Roman E. Maeder. Object-oriented programming. *The Mathematica Journal*, 3(1), 1993.

[31] Roman E. Maeder. Uniform polyhedra. *The Mathematica Journal*, 3(4), 1993.

[32] Benoit B. Mandelbrot. *Fractals: Form, Chance, and Dimension*. W. H. Freeman, 1977.

[33] Benoit B. Mandelbrot. *The Fractal Geometry of Nature*. W. H. Freeman, 1982.

[34] John McCarthy. Recursive functions of symbolic expressions and their computation by machine I. *J. of the ACM*, 3:184–195, 1960.

[35] Michael McGuire. *An Eye for Fractals*. Addison-Wesley, 1991.

[36] Bertrand Meyer. *Object-Oriented Software Construction*. Prentice Hall, 1988.

[37] Arthur Ogawa. The trinoid revisited. *The Mathematica Journal*, 2(1), 1992.

[38] Robert Osserman. *A Survey of Minimal Surfaces*. Van Nostrand Reinhold Company, 1969.

[39] Oren Patashnik. *BIBTEXing*, February 8, 1988.

[40] Heinz-Otto Peitgen and Dietmar Saupe. *The Science of Fractal Images*. Springer-Verlag, 1988.

[41] Tomas Rokicki. *DVIPS: A TeX Driver*, 1993.

[42] Dana S. Scott. Data types as lattices. *SIAM J. of Computing*, 5:522–587, 1976.

[43] J. E. Stoy. Foundations of denotational semantics. In D. Bjorner, editor, *Abstract Software Specification*, volume 86 of *SLNCS*. Springer-Verlag, New York, 1980.

[44] Alan M. Turing. On computable numbers with an application to the Entscheidungsproblem. *P. Lond. Math. Soc. (2)*, 42:230–265, 1936-7.

[45] Jeffrey D. Ullman. *Principles of Database and Knowledge-base Systems*, volume 14 of *Principles of Computer Science Series*. Computer Science Press, Rockville, Maryland, 1988.

[46] Jan van Leeuwen, editor. *Formal Models and Semantics*, volume B of *Handbook of Theoretical Computer Science*. Elsevier, 1990.

[47] H. von Koch. Sur une courbe continue sans tangente, obtenue par une construction geometrique elementaire. *Arkiv for Matematik, Astronomi och Fysik*, 1:681–704, 1904.

[48] Martin Wirsing. Algebraic specification. In van Leeuwen [46].

Additional Program Listings

This section of the appendix contains listings of programs not reproduced in the main text of the book. Most of these are packages used to produce some of the illustrations. All of these programs are included in the floppy disk.

Pictures.m

The package Pictures.m is used to produce the illustrations for the chapter openers. The floppy disk includes a notebook version, Pictures.ma. The pictures are rendered by evaluating the symbol chapter$n$ for chapter no. n. The initialization sets up all definitions but doesn't render the pictures yet.

```
Needs["Graphics`Animation`"]
Needs["Statistics`DataManipulation`"]
Needs["Graphics`Graphics`"]

<< FlipBookAnimation.m

Share[]

Needs["MinimalSurfaces`"]
Needs["UniformPolyhedra`"]
Needs["Spirals`"]
Needs["FractCube`"]
Needs["Icosahedra`"]
Needs["Tetrix`"]
Needs["RandomWalk3D`"]
Needs["fBm`"]

<< FractalExamples.m

Share[]

(* Wythoff symbols *)
poly = {
    {w1[4, 2, 3],   w1[3, 2, 4],   w1[2, 3, 4]},
    {w1[5, 2, 3],   w1[3, 2, 5],   w1[2, 3, 5]},
    {w1[5, 2, 5/2], w1[5/2, 2, 5], w1[2, 5/2, 5]},
    {w1[3, 2, 5/2], w1[5/2, 2, 3], w1[2, 5/2, 3]}
};

chapter1 :=
    Show[ GraphicsArray[ Map[Graphics3D[MakeUniform[#], Boxed->False]&, poly, {2}],
                        GraphicsSpacing->0 ]
    ]

icosahedra = {{3, 11}, {20, 36}}
```

```
chapter2 :=
    Show[ GraphicsArray[ Map[ Stellation[#, Boxed->False]&, icosahedra, {2} ],
                         GraphicsSpacing->0 ]
    ]

chapter3 := PrimesPlot[ 199 ]

chapter4 := Show[ RandomWalk3D[1000] ]

chapter5 :=
    Show[ Tetrix[4, Boxed->False],
          LightSources -> {{{1., 0., 5.}, RGBColor[1, 0, 0]},
                           {{1., 1., 1.}, RGBColor[0, 1, 0]},
                           {{0, 1., 0.4}, RGBColor[0, 0, 1]}},
          ViewPoint->{1.4, -2.4, 0.9}
    ]

chapter6 :=
    Show[ GraphicsArray[ Partition[ FractalCube[4, #]& /@ Range[0.1, 0.9, 0.1], 3],
                         GraphicsSpacing->0 ]
    ]

chapter7 :=
    Show[ ContourGraphics[fBm2D[128, 3 - 2.4, 64]],
          Contours -> 20, FrameTicks -> None ]

chapter8 := Show[ Flowsnake[5], Thickness->0.001 ]

chapter9 :=
    Animate[ CartesianSurface[Exp[I arg] E^-z, E^z, z,
                  {-2, 2, 4/20}, {-Pi, Pi, Pi/14},
                  PlotRange->{{-7.5, 7.5}, {-7.5, 7.5}, {-7.5, 7.5}},
                  Boxed->False, Axes->None],
             {arg, 0, Pi/2},
             Frames->9, Closed->False ];

system = Names["System`**"];
initials = First /@ Characters /@ system;
initials = ToUpperCase /@ initials;
all = Union[ToUpperCase /@ Flatten[$Letters]]; (* all letters *)
freq = Frequencies[Join[all, initials]];
freq = Apply[{#1-1, #2}&, freq, {1}];   (* discount one for "all" *)
chapter10 := BarChart[freq];
```

Gallery.m

The package Gallery.m contains the code for those color plates that were produced directly from *Mathematica*. The remaining color pictures used data from *Mathematica* for photorealistic rendering on special hardware. For these, the code below will produce only an approximation of the pictures reproduced in the color insert.

```
Needs["Graphics`Colors`"]

Needs["Spirals`"]
Needs["MinimalSurfaces`"]
Needs["UniformPolyhedra`"]
Needs["ComplexPlot3D`"]
Needs["Sierpinski`"]
Needs["Icosahedra`"]
Needs["fBm`"]

Share[]


plate01 := DivisorPlot[199]

plate02a :=
    PolarSurface[ z, 1/z, z, {0.00005, 2.0005}, {-Pi, 7Pi},
                  ViewPoint -> {-2.1, -1.1, 1.2}, Axes -> None,
                  PlotRange -> All, PlotPoints -> {12, 100}
    ]

plate02b :=
    PolarSurface[ 1, z, z, {0, 2.5}, Axes -> None,
                  PlotRange -> All, PlotPoints -> {16, 60}
    ]

plate02c :=
    PolarSurface[ 2-2/z^4, z, z, {0.35, 0.85}, {-Pi, Pi},
                  Axes -> None, PlotRange->All, PlotPoints->{16, 80}
    ]

plate03a := Show[ Graphics3D[ MakeUniform[w1[2, 5/2, 3]] ] ];
plate03c := Show[ Graphics3D[ MakeUniform[w1[5, 2, 5/2]] ] ];
plate03d := Show[ Graphics3D[ MakeUniform[w1[5/2, 2, 5]] ] ];
plate03e := Show[ Graphics3D[ MakeUniform[w1[3, 2, 5/2]] ] ];
plate03f := Show[ Graphics3D[ MakeUniform[w1[5/2, 2, 3]] ] ];


plate04a :=
    ComplexPlot3D[ Sin[z], {z, -Pi - I, Pi + I}, PlotPoints->50 ]
plate04b :=
    ComplexPlot3D[ Exp[z], {z, -1 - 4I, 1 + 4I}, PlotPoints->50 ]
plate04c :=
    ComplexPlot3D[ Exp[1/z], {z, -0.4 - 0.4I, 0.4 + 0.4I}, PlotPoints->120,
        PlotRange->{0, 500}, ClipFill->None, BoxRatios->{1,1,1.2}, Mesh->False,
        AxesEdge->{{-1,-1}, {-1,-1}, {-1, 1}}, ViewPoint->{-2.4, -1.3, 2} ]
plate04d :=
    ComplexPlot3D[ 1/z^2, {z, -0.4 - 0.4I, 0.4 + 0.4I}, PlotPoints->120,
        PlotRange->{0, 500}, ClipFill->None, BoxRatios->{1,1,1.2}, Mesh->False,
        AxesEdge->{{-1,-1}, {-1,-1}, {-1, 1}}, ViewPoint->{-2.4, -1.3, 2} ]

plate05 :=
    Show[ Sierpinski[3],
          ViewPoint->{0.95, -3.1, 0.8},
          LightSources -> {{{1., 0., 5.}, RGBColor[1, 0, 0]},
                           {{1., 1., 1.}, RGBColor[0, 1, 0]},
```

```
                    {{0., 1., 0.4}, RGBColor[0, 0, 1]},
                    {{0., -1., 0.4}, RGBColor[0, 0, 1]}},
          Boxed->False, Background -> GrayLevel[0]
    ]

icosa = {
    { 2,  3,  5},
    {12,  7, 20},
    {32, 36, 11},
    {52, 59,  4}
};
plate06 :=
    Show[ GraphicsArray[ Map[ Stellation[#, Boxed->False]&, icosa, {2} ],
                    GraphicsSpacing->0 ]
    ]

plate07 := plate07 =
    Show[ fBm2Surface[0.1 fBm2D[128, 3-#, 64], MeshStyle->Thickness[0]],
          ViewPoint->{1.3, -2.4, 0.8} ]& /@
    Range[2.2, 2.8, 0.2]
makeSea[sg_SurfaceGraphics] :=
    Module[{max},
        max = FullOptions[sg, PlotRange][[3, 2]];
        Show[ sg, Mesh->False, BoxRatios->{1,1,0.15},
              PlotRange->{0, 1.1 max}, ClipFill->SkyBlue,
              ViewPoint->{1.3, -2.4, 0.6} ]
    ]
plate08 := makeSea /@ plate07
```

fBm.m

Fractional Brownian motion is a good model for natural looking fractal scenes. The function
$fBm2D[n, H]$ generates a fractal mountain landscape with dimension $3 - H$. Images for various H are shown in plates 7 and 8. The source for this program is the book *The Sciene of Fractal Images* by Peitgen and Saupe [40].

```
BeginPackage["fBm`", "Statistics`NormalDistribution`"]

fBm1D::usage = "fBm1D[n, H, (m)] generates a fractional Brownian motion of length n
    with dimension 2 - H, 0 < H < 1. m limits the higher frequencies to include."
fBm2D::usage = "fBm2D[n, H, (m)] generates a fractional Brownian surface of size n^2
    with dimension 3 - H, 0 < H < 1. m limits the higher frequencies to include."
fBm2Surface::usage = "fBm2Surface[mesh, opts...] turns the output of fBm2D into a
    surface graphics object."

Begin["`Private`"]
```

```mathematica
pi2 = N[2Pi]

gauss := Random[NormalDistribution[ ]]
randPhase := Random[Real, {0, pi2}]

(* one dimension, program SpectralSynthesisFM1D *)

(* fractal dimension is 2 - H *)

fBm1D[n_?EvenQ, H_, maxn_:Automatic] :=
    Module[{ff, n2 = n/2, beta = -(2H + 1)/2, i, nmax = maxn},
        If[ nmax === Automatic, nmax = n ];
        nmax /= 2;
        ff = Table[ freq1[i, beta, nmax], {i, 0, n2} ];
        ff[[-1]] = Re[ff[[-1]]];
        ff = Join[ ff, Reverse[Conjugate[Take[ff, {2, -2}]]] ];
        n Chop[InverseFourier[ff]]
    ]

freq1[0, exp_, nmax_] := 0
freq1[i_, exp_, nmax_] := If[i > nmax, 0, gauss i^exp Exp[I randPhase]]


(* two-dimensional, program SpectralSynthesisFM2D *)

(* fractal dimension is 3 - H, 0 < H < 1 *)

fBm2D[n_?EvenQ, H_, maxn_:Automatic] :=
    Module[{line0, half1, linem, half2, full,
            n2 = n/2, beta = -(H + 1)/2, i, j, nmax = maxn},
        If[ nmax === Automatic, nmax = n ];
        nmax /= 2;
        line0 = Table[freq2[0, j, beta, nmax], {j, 0, n2}];
        line0[[-1]] = Re[line0[[-1]]];
        line0 = Join[ line0, Conjugate[Reverse[Take[line0, {2, -2}]]] ];

        half1 = Table[ freq2[i, Min[j, n-j], beta, nmax], {i, 1, n2-1}, {j, 0, n-1} ];

        linem = Table[freq2[n2, j, beta, nmax], {j, 0, n2}];
        linem[[1]] = Re[linem[[1]]];
        linem[[-1]] = Re[linem[[-1]]];
        linem = Join[ linem, Conjugate[Reverse[Take[linem, {2, -2}]]] ];

        half2 = RotateLeft /@ half1;
        half2 = Reverse[ Reverse /@ half2 ];
        half2 = Conjugate[half2];

        full = Join[ {line0}, half1, {linem}, half2 ];
        full = N[Sqrt[n]] Chop[InverseFourier[full]];

        full
    ]

freq2[0, 0, exp_, nmax_] := 0
freq2[i_, j_, exp_, nmax_] :=
    If[i > nmax || j > nmax, 0, gauss (i^2+j^2)^exp Exp[I randPhase]]
```

```
fBm2Surface[vals_List, opts___] :=
    SurfaceGraphics[vals, {opts,
        MeshRange->0.5{{-1, 1}, {-1, 1}}, BoxRatios->Automatic} ]
End[ ]

EndPackage[ ]
```

FractCube.m

This small piece of code produced the illustration for Chapter 6.

```
BeginPackage["FractCube`"]

FractalCube::usage = "FractalCube[n, prob, opts...] renders a cube with n binary
    subdivision taken with probability prob."
cheap::usage = "cheap = True renders only the faces visible from a positive viewpoint."

Begin["`Private`"]

`randBit
cheap = False;

(* render only three faces *)

cube[0, {x_, y_, z_}]/; cheap := {
    Polygon[{{x+1, y, z}, {x+1, y, z+1}, {x+1, y+1, z+1}, {x+1, y+1, z}}],
    Polygon[{{x+1, y+1, z}, {x+1, y+1, z+1}, {x, y+1, z+1}, {x, y+1, z}}],
    Polygon[{{x+1, y+1, z+1}, {x+1, y, z+1}, {x, y, z+1}, {x, y+1, z+1}}]
}

(* render all faces *)

cube[0, lll_] := {Cuboid[lll]}

cube[n_, {x0_, y0_, z0_}] :=
  With[{s = 2.0^(n-1), n1 = n-1},
    With[{ lll = Outer[List, {x0, x0 + s}, {y0, y0 + s}, {z0, z0 + s}],
           rand = Table[randBit, {2}, {2}, {2}]},
      MapThread[ If[#1, cube[n1, #2], {}]&, {rand, lll}, 3 ]
  ]]

FractalCube[n_Integer?NonNegative, prob_Real, opts___] :=
    Module[{polylist, s = 2.0^(n-1)},
        randBit := Random[] < prob;
        polylist =  cube[n, {-s,-s,-s}];
    polylist = Flatten[ polylist ];
    Graphics3D[ polylist, opts,
                PlotRange->1.01{{-s,s}, {-s,s}, {-s,s}},
                ViewPoint -> {1.3, 2.4, 2.} ]
    ]
End[ ]

EndPackage[ ]
```

ComplexPlot3D.m

Code for displaying complex-valued functions used for plate 4.

```
(* 3D Plots in the Complex Plane with Color *)

(* by Kevin McIsaac and Roman Maeder *)

BeginPackage["ComplexPlot3D`", "Graphics`ArgColors`"]

ComplexPlot3D::usage = "ComplexPlot3D[fz, {z, z0, z1}, (options)]
    Plots fz in a rectangle in the complex plane.
    The absolute value is represented by the height
    and phase is represented by the hue."

Begin["`Private`"]

ComplexPlot3D[fz_, {z_, z0_, z1_}, opts___ ] :=
    Module[{fval, x, y},
        Plot3D[{Abs[fval = fz /. z -> x + I y], ArgColor[fval]},
                {x, Re[z0], Re[z1]}, {y, Im[z0], Im[z1]}, opts]
    ]

End[ ]

Protect[ComplexPlot3D]

EndPackage[ ]
```

RandomWalk3D.m

A three-dimensional random walk was shown at the beginning of Chapter 4.

```
BeginPackage["RandomWalk3D`"]

RandomWalk3D::usage = "RandomWalk3D[n, opts] is a random walk in 3D."

Begin["`Private`"]

units = Flatten[Outer[ #2 RotateRight[{1, 0, 0}, #1]&, {0, 1, 2}, {-1, 1} ], 1]

RandomDir[ ] := units[[ Random[Integer, {1, Length[units]}] ]]

RandomWalk3D[n_, opts___] :=
    Graphics3D[ Cuboid /@ Union[ NestList[ # + RandomDir[]&, {0,0,0}, n ] ],
                {opts} ]

End[ ]

EndPackage[ ]
```

Sierpinski.m

This is the code for the Sierpinski sponge. The picture appeared first in *Programming in Mathematica* and is given in a ray traced version in plate 5. The code is similar to Tetrix.m above.

```
BeginPackage["Sierpinski`"]

Sierpinski::usage = "Sierpinski[n, opts...] is the n-th iteration of the
    Sierpinski sponge. Only viewpoints in the default octant are supported"

Begin["`Private`"]

cheese[0, False, False, False, __ ] := {} (* small optimization *)

cheese[0, right_, front_, top_, x0_, y0_, z0_] :=
  With[{xs = x0+1, ys = y0+1, zs = z0+1},
    { If[right,
        Polygon[{{xs, y0, z0}, {xs, ys, z0}, {xs, ys, zs}, {xs, y0, zs}}],
        {} ],
      If[front,
        Polygon[{{x0, y0, z0}, {xs, y0, z0}, {xs, y0, zs}, {x0, y0, zs}}],
        {} ],
      If[top,
        Polygon[{{x0, y0, zs}, {xs, y0, zs}, {xs, ys, zs}, {x0, ys, zs}}],
        {} ]
    }
  ]

cheese[n_, right_, front_, top_, x0_, y0_, z0_] :=
  With[{ xs = x0 + 3^(n-1), xt = x0 + 2 3^(n-1),
         ys = y0 + 3^(n-1), yt = y0 + 2 3^(n-1),
         zs = z0 + 3^(n-1), zt = z0 + 2 3^(n-1),
         n1 = n - 1},
    { (* bottom layer *)
    cheese[n1, False, front, False, x0, y0, z0],
    cheese[n1, False, front, True , xs, y0, z0],
    cheese[n1, right, front, False, xt, y0, z0],
    cheese[n1, True , False, True , x0, ys, z0],
    cheese[n1, right, False, True , xt, ys, z0],
    cheese[n1, False, False, False, x0, yt, z0],
    cheese[n1, False, True , True , xs, yt, z0],
    cheese[n1, right, False, False, xt, yt, z0],
      (* middle layer *)
    cheese[n1, True , front, False, x0, y0, zs],
    cheese[n1, right, front, False, xt, y0, zs],
    cheese[n1, True , True , False, x0, yt, zs],
    cheese[n1, right, True , False, xt, yt, zs],
      (* top layer *)
    cheese[n1, False, front, top  , x0, y0, zt],
    cheese[n1, False, front, top  , xs, y0, zt],
    cheese[n1, right, front, top  , xt, y0, zt],
    cheese[n1, True , False, top  , x0, ys, zt],
    cheese[n1, right, False, top  , xt, ys, zt],
    cheese[n1, False, False, top  , x0, yt, zt],
    cheese[n1, False, True , top  , xs, yt, zt],
    cheese[n1, right, False, top  , xt, yt, zt]
    }
  ]
```

```
Sierpinski[ n_Integer?NonNegative, opts___ ] :=
    Module[{polylist},
        polylist = {EdgeForm[], cheese[n, True, True, True, 0, 0, 0]};
        polylist = Flatten[ polylist ];
        Graphics3D[ polylist, opts, Boxed->False ]
    ]

End[]

EndPackage[]
```

Spirals.m

A spiral plot of the location of prime numbers was shown at the beginning of Chapter 3. A color illustration shown the number of divisors of integers is given on plate 1.

```
BeginPackage["Spirals`", "Graphics`ArgColors`"]

Spiral::usage = "Spiral[f, n] is an n by n table of values f[1] .. f[n^2]
    arranged in a spiral. n must be odd."
Wind::usage = "Wind[list] winds the elements of list in a spiral."

SpiralGraphics::usage = "SpiralGraphics[spiral, opts...] shows a table of
    gray values (as Raster[])."
SpiralColorGraphics::usage = "SpiralColorGraphics[spiral, opts...] shows a table of
    color directives (as RasterArray[])."

PrimesPlot::usage = "PrimesPlot[n] shows the primes among 1..n^2 in a spiral."
DivisorPlot::usage = "DivisorPlot[n] shows the number of divisors of 1..n^2
    in a color spiral."

Begin["`Private`"]

Spiral[f_, n_?OddQ] :=
    Module[{x = (n+1)/2, y = (n+1)/2, i = 0, ring, t = Table[Null, {n}, {n}]},
        t[[y, x]] = With[{i=++i}, f[i]];
        y += 1;
        Do[
            Do[ t[[y, x]] = With[{i=++i}, f[i]]; x -= 1, {ring} ];
            {x, y} += {1, -1};
            Do[ t[[y, x]] = With[{i=++i}, f[i]]; y -= 1, {ring} ];
            {x, y} += {1, 1};
            Do[ t[[y, x]] = With[{i=++i}, f[i]]; x += 1, {ring} ];
            {x, y} += {-1, 1};
            Do[ t[[y, x]] = With[{i=++i}, f[i]]; y += 1, {ring} ],
        {ring, 2, n-1, 2} ];
        t
    ]

SpiralGraphics[spiral_, opts___] :=
    Graphics[ Raster[Reverse[spiral]],
            {opts, AspectRatio->1, Frame->True, FrameTicks->None} ]
```

```
SpiralColorGraphics[spiral_, opts___] :=
    Graphics[ RasterArray[Reverse[spiral]],
              {opts, AspectRatio->1, Frame->True, FrameTicks->None} ]

Wind[l_List] :=
    With[{n = Floor[N[Sqrt[Length[l]]]]},
        Spiral[ l[[#]]&, 2Floor[(n+1)/2]-1 ]
    ]

PrimesPlot[n_, opts___] :=
    Show[ SpiralGraphics[Spiral[If[PrimeQ[#], 0, 1]&, n], opts] ]

DivisorPlot[n_, opts___] :=
    Module[{maxd, div, pi2 = N[2Pi]},
        div = (# - EulerPhi[#])& /@ Range[n^2];
        maxd = Max[div] + 1;
        div = pi2 div/maxd;
        spiral = Wind[ ColorCircle /@ div ];
        Show[ SpiralColorGraphics[ spiral, opts ] ]
    ]

End[]

EndPackage[]
```

Tetrix.m

The tetrahedral sponge is displayed in the opening of Chapter 5.

```
BeginPackage["Tetrix`"]

Tetrix::usage = "Tetrix[n, opts...] is the n-th iteration of the tetrahedral
    fractal sponge."

Begin["`Private`"]

(* edge directions *)

e1 = N[{2^(1/2), (2^(1/2)*3^(1/2))/3, (4*3^(1/2))/3}]
e2 = N[{2^(1/2), 2^(1/2)*3^(1/2), 0}]
e3 = N[{2*2^(1/2), 0, 0}]

cheese[0, p_, s_] :=
    With[{p1 = p + s e1, p2 = p + s e2, p3 = p + s e3},
        {Polygon[{p , p1, p2}],
         Polygon[{p , p2, p3}],
         Polygon[{p , p3, p1}],
         Polygon[{p1, p2, p3}]}
    ]
```

```
cheese[n_, p_, size_] :=
    With[{s = size/2, n1 = n-1},
        {
            cheese[n1, p, s],
            cheese[n1, p + s e1, s],
            cheese[n1, p + s e2, s],
            cheese[n1, p + s e3, s]
        }
    ]

Tetrix[ n_Integer?NonNegative, opts___ ] :=
    Graphics3D[ cheese[n, {-1.0, -1.0, -1.0}, 2.0], {opts} ]

End[ ]

EndPackage[ ]
```

Index of Programs

This table shows the page numbers where the programs are listed or explained in detail.

All these programs are part of the enclosed floppy disk. Additionally, it contains the following packages and notebooks, which are not listed in the book: FractalExamples.m, FractalExamples.ma, Gallery.ma, Icosahedra.m, MinimalExamples.ma, Pictures.ma, RecordManager.m, StreamExamples.ma, UniformPolyhedra.m, and music.db.

Index

Symbols beginning with a dollar sign $ are listed under the following letter. Names of persons are given in italics. Further typographical conventions are explained on page xv.